Routledge Revivals

Science at the Cross Roads

The papers given by the Soviet Delegation to the Second International Congress of the History of Science and Technology in London in 1931, headed by N. I. Bukharin, exerted a profound influence on Western historiography of science. Perhaps the most influential contribution was that of Hessen, who made a long and classical statement of Marxist historiography, taking Isaac Newton as his example. The collection, which appeared in Britain at the height of the Depression, fostered an acute social awareness and a heated debate among many working scientists. Accredited by some as "the starting point of a new evaluation of the history of science", the book reflects the huge social and economic divide between Socialism and Capitalism present at the time of publication, and its influence on intellectual culture and scientific advancement.

Science at the Cross Roads

Papers from the Second International Congress of the History of Science and Technology 1931

N. I. Bukharin et al.

Routledge
Taylor & Francis Group

First published in 1931
Second edition 1971
by Frank Cass and Company Limited

This edition first published in 2013 by Routledge
2 Park Square, Milton Park, Abingdon, Oxon, OX14 4RN

Simultaneously published in the USA and Canada
by Routledge
711 Third Avenue, New York, NY 10017

Routledge is an imprint of the Taylor & Francis Group, an informa business

Publisher's Note
The publisher has gone to great lengths to ensure the quality of this reprint but points out that some imperfections in the original copies may be apparent.

Disclaimer
The publisher has made every effort to trace copyright holders and welcomes correspondence from those they have been unable to contact.

A Library of Congress record exists under LC control number: 74171262

ISBN 13: 978-0-415-82513-9 (hbk)
ISBN 13: 978-0-203-38393-3 (ebk)

Published by
FRANK CASS AND COMPANY LIMITED
67 Great Russell Street, London WC1B 3BT

First edition	1931
Second edition	1971

ISBN 0 7146 2868 9

Printed in Great Britain by
Biddles Ltd., Guildford, Surrey

EDITOR'S NOTE

Forty years ago this year, the IInd International Congress of the History of Science and Technology was held in London. The Congress of 1931 was overshadowed by the world-wide crisis. It could easily have passed unnoticed, yet it was destined to become a milestone in the historiography of science, and to make a profound and indelible impression upon some of its most promising young participants. The unexpected appearance of a large delegation from the USSR presented Western historians, many for the first time, with a sustained Marxist treatment of social and economic factors as elements in scientific and technological development. The collective impact of these views, presented in *Science at the Cross Roads,* launched an historiographical debate which in many respects remains relevant today.

For many years this volume has been out of print and virtually unobtainable. The occasion of the XIIIth Congress of the History of Science and Technology in Moscow in August 1971 has provided a natural opportunity for reprinting and reflecting upon this influential work "forty years on". By way of introduction we have brought together the reflections of one of the most distinguished participants of the Congress, and the fresh research and interpretations of a young historian. Today, as the historiography of science is undergoing close reappraisal, the study of science in its social context is compelling wide international attention. Readers of *Science at the Cross Roads* in 1971 will judge for themselves the strengths and weaknesses of the arguments presented. But it is within this context of reappraisal that the Essays of 1931 have still an instructive role to play.

FOREWORD

by
Joseph Needham, F.R.S.

Master of Caius College, Cambridge

In this volume a group of contributions presented to the IInd International Congress of the History of Science, held at London in 1931, by a Soviet delegation headed by N. Bukharin, again—deservedly—sees the light. Published together at the time by the Russian Foreign-Languages Press, Kniga, under the title "Science at the Cross Roads", they circulated for some time in book form, but the volume soon went out of print and has long been very scarce. As one of the few remaining members of that congress I am pleased to write a foreword for this reprint.

The Congress occurred at a particularly important time for me. The year 1931 saw the publication of my three-volume "Chemical Embryology", the first volume of which was partly taken up by a study of the history of embryology down to the beginning of the nineteenth century. This was intended to preface my account of the incursion of biochemistry into the morphological and developmental sciences. In working on this historical introduction I had got to know Charles and Dorothea Singer, and my wife and I had started that long friendship with them which lasted down to their deaths in recent years. We often used to spend week-ends or whole weeks with them in London and later at Kilmarth in Cornwall. I was thus able to make use of Charles Singer's wonderful personal library in its romantic setting overlooking St. Austell Bay, and always tried to acknowledge my debt to him—as for example in the dedication to the separately published "History of Embryology", which came out in 1934. It was therefore natural enough for me to attend the International Congress in London in 1931; and having for a long time already been on the left politically I was very ready to give a sympathetic hearing to the Russian delegation—unexpectedly heavy-weight in character—which made its appearance at the Congress. The same perhaps could hardly be said of all the other members, and one of my most vivid recollections is of Charles Singer in the Presidential chair, trying to shut the Russians up, when they had had their twenty minutes, by the aid of a large ship's bell which he continuously rang. They, of course, had come expecting to be able to speak for hours.

Perhaps the outstanding Russian contribution was that of Boris Hessen, who made a long and classical statement of the Marxist historiography of science, taking as his subject of analysis Isaac Newton. Here was a paradigm of the traditional history of science, so great a genius that he could not have been influenced by his environment at all, and certainly not by a sub-conscious appreciation of the needs of the society of the rising bourgeoisie of the seventeenth century. To suggest such a thing was, in terms of conventional thinking, almost a sacrilegious act, in any case culpable of lèse-majesté. Yet this was the case Hessen worked out *in extenso*, tripping over proper names and making mistakes of detail on the way, but producing a veritable manifesto of the Marxist form of externalism in the history of science. Newton had not, it seemed, lived his life in a vacuum, he had been aware of the practical needs of the early capitalist society of his time, and the subjects in which he interested himself, however far they took him among the stars and planets, or to the confines of theological thought, were in fact those where solutions and new starts were needed—for applied mathematics, hydrodynamics, navigation, ballistics, mechanics, metallurgy and the like. This essay, with all its unsophisticated bluntness, had a great influence during the subsequent forty years, an influence still perhaps not yet exhausted; hence its present reprinting is to be welcomed. Whether or not I was actually one of the audience when Hessen gave his lecture I cannot clearly remember, though I think I was; at all events he must have been suppressed by the bell before he had got very far with his epoch-making text.

The debate between externalists and internalists will go on, no doubt, for many a long year yet, with those who feel they can descry profound influences of social structure and social change upon science and scientific thought opposing those who prefer to think only in terms of an internal logic of development powered by intellectual giants of mysterious origin. With the appearance on the scene of intensive studies of mathematics, science, technology and medicine in the great non-European civilisations the debate is likely to sharpen, for the failure of China and India to give rise to distinctively modern science while having been ahead of Europe for fourteen previous centuries is going to take some explaining. Having worked in these fields for the past thirty years myself, it seems to me that the internalist doctrine is likely to encounter grave difficulties, because the "ideological superstructure", or, in ordinary English, the intellectual, philosophical, theological and cultural systems of ideas of the Asian civilisations are not going to be able to take the causal weight required. Some of these idea-systems, in fact, such as Taoism and Neo-Confucianism, would seem to have been much more congruent with modern science than any of the European ones were, notably Christian theology. It will therefore fall to the externalists to investigate all the social and economic characteristics of the Chinese, Indian, Arabic and European civilisations, and to see how far, for example, the inhibition of the rise of the merchant class to State power, might take us in explaining how Galilean science could come to birth in Pisa but not in Patna or Peking. The trumpet-blast of Hessen may therefore still have great value in orienting the minds of younger scholars towards a direction fruitful for historical analyses still to come, and may lead in the end to a deeper understanding of the mainsprings and hindrances of science in East and West, far more subtle and sophisticated than he himself could ever hope to be.

Besides the Hessen paper a number of contributions, here also reproduced, merit attention. The introduction by N. Bukharin was also in its way a classical statement of the Marxist position, and the memorable essay on the origins of Old-World agriculture by the great N. I. Vavilov retains all its interest. Influential, too, on theoretical biologists in the thirties, was B. Zavadowsky's contribution directed against "reductionism" and in support of non-obscurantist organicism applied to the successive integrative levels of living things. "The true task of scientific research", he wrote, "is not the violent identification of the biological and the physical, but the ability to discover the qualitatively specific controlling principles which characterise the principal features of every given phenomenon, and to find methods of research appropriate to it." In other words, each level of complexity and organisation, physical, chemical, biological, sociological, has to be studied at the relevant level, and the appropriate regularities found there—then only can meaning be introduced into the whole by scrutinising the relations between the levels. For example, our knowledge of the laws of Mendelian inheritance did not have to await the elucidation of the nucleoprotein molecule, though our understanding of the whole is immensely increased thereby. Lastly, E. Kol'man's discussion was of great interest to mathematicians.

It would hardly be possible in a foreword such as this to ignore the tragic fact of the disappearance of so many of these delegates in the years after the Congress, according to the dreadful principle that "all revolutions devour their own children." Of Bukharin as a famous political leader there is no need to speak, but the Hamlet of the piece, B. Hessen, published hardly anything more, so far as we know, after the London meeting, and one presumes that he fell a victim to the Stalinist "illegalities". This we know to have been the case with Vavilov the geneticist, mown down in the, for a time, triumphant advance of Lysenkoism. All the more delighted was I, therefore, to meet again Professor Kol'man at the XIth International Congress of the History of Science at Warsaw and Cracow in 1965—safe and sound, but having suffered many years of danger and imprisonment.

On this note, therefore, with warm greetings to all our Russian colleagues of today, and best memories of all that their predecessors contributed to the IInd International Congress, I may end these few words and commend the reprint to its many readers.

INTRODUCTION

On the Reception of *Science At The Cross Roads* in England

by

P. G. Werskey

Lecturer, Science Studies Unit, University of Edinburgh

Science at the Cross Roads contains the contributions of the Soviet delegation to the Second International Congress of the History of Science and Technology, held in London between June 29th and July 4th, 1931. The decision to produce this book was made immediately after the announcement on the first day of the Congress that a special Saturday morning session (July 4th) would be set aside for presentation of the Russians' papers. In the ensuing five days, a battery of delegates, translators and proofreaders laboured feverishly at the Soviet Embassy on the proposed volume. Although there was not enough time to bind the collected articles together, they were made available as separate units for the meeting on Saturday. *Science at the Cross Roads* itself appeared ten days later in a limited edition, bearing ample testimony in the form of unidiomatic English phrases, typographical errors and transposed lines to the haste which had accompanied its creation.[1]

Despite the book's unprofessional appearance, there is no doubt that it is a cardinal document in the history of English and Russian Marxism. On the Soviet side, it represents the collective position of an important group of administrators, philosophers and scientists on the philosophy and politics of science during a critical period of 'the great break'.[2] Ironically, many of their intellectual perspectives were to be extended most fruitfully not in their native land, but in Great Britain by a small number of young Marxist academics. Within this latter group, it was the natural scientists who most eagerly embraced the message(s) contained in *Science at the Cross Roads*. For them, the book not only marked "the starting point of a new evaluation of the history of science,"[3] but also demonstrated the impossibility of utilising science for social reconstruction "within the framework of a chaotic capitalism."[4] Their political and philosophical opponents in England were thus able to deduce later from such statements that "the movement against pure science and against freedom in science was first brought to Great Britain by the Soviet delegation to the International Congress of the History of Science and Technology held in London in 1931."[5]

The reprinting of *Science at the Cross Roads* in the Cass Library of Science and Public Affairs provides an appropriate occasion for its re-evaluation. The essay which follows will stress the circumstances surrounding the Russians' success at the Congress and the interaction between the British and Soviet delegates in London, and will conclude with a brief description of the fate of 'Marxist science' in Britain since 1931.

The Right Place at the Right Time

While all sides have agreed that the Russians did make a considerable impact in London during the Congress, no one has offered a satisfactory explanation as to why this was the case. Anti-Marxists have argued that, given the economic depression, "attention in 1931 was naturally focused on economic matters, and this preoccupation lent impetus to the specifically Marxist doctrine . . . that all scientific progress was really determined by economic causes".[6] This rationalization, however, is both too sweeping and too narrow in its scope, in that it allows us neither to account for the vast majority of British scientists who lived through the Depression without ever considering the Marxist world-view, nor to understand the complex factors which led some researchers to Marxism *long before the mid-summer meeting of* 1931. Needless to say, the left-wing scientists were well aware of the latter point. As Hyman Levy later suggested, "the standpoint consistently adopted by these [Russian] delegates crystallized out in remarkable fashion what had been simmering in the minds of many for some time past."[7] The one truly revealing paper was that of Hessen on "The Social and Economic Roots of Newton's 'Principia' ", which attempted to relate the scientific revolution of the seventeenth century to the rise of bourgeois capitalism. But, having said this much, the young Marxists appear to have assumed that the Russian contributions embodied both a long-standing philosophical tradition and the official blueprint for Soviet scientific development in the future. Recent scholarship has failed to substantiate either of these assumptions.[8] Our analysis must therefore begin with a consideration of the position of the Russian scientific community on the eve of the Congress.

The period between 1929 and 1932 has been described as the 'great break' in the relationship between scientists and the government in the Soviet Union. Prior to that time, few political constraints were forced upon natural scientists, whose services as 'bourgeois specialists' were recognised as critical to the earliest phases of the Revolution. The fact that the Academy of Sciences was the last Tsarist body to be reformed (1930) was symbolic of the institutional autonomy which Russian science by and large retained in the Twenties. Certainly there were, at that time, many proposals for both the 'proletarianisation' of science and its transformation along dialectical lines. But it was the 'Deborinite' school of philosophers, with its emphasis on the dialectical quality of contemporary science (relativity theory, 'Morganist' genetics, etc.), which ultimately enjoyed the Central Committee's blessing, at least up until 1930.[9]

The period of grace for Soviet scientists was swiftly brought to an end, however, with Stalin's consolidation of power in 1929. Shortly thereafter, the Academy was 'bolshevized' through the election of Communist party members and politically loyal, industrial scientists and engineers (previously excluded by Academicians oriented towards basic research); a few months later the institution received a new Charter which made it an official part of the State bureaucracy.[10] 'Bolshevization' also took the form of renewed attacks by Party philosophers against 'bourgeois' science, demands for the expression of political loyalty on the part of older scientists and recruitment of talented peasants and workers into the ranks of the scientific élite. Thus, for the first time, academic researchers were obliged to defend their work in detail against the charge that it was either anti-Marxist or irrelevant to the achievement of agricultural and industrial goals set out in the first Five-Year Plan. The net effect of such developments was to intensify conscious discussion of the philosophical bases of science, as well as the relationship of research to national life, among officials and scientists alike.[11]

For a state founded on the principles of Marxism-Leninism, it was inevitable that 'science' would play a prominent role in political affairs. The distinction between "scientific" and "utopian" socialism lay at the heart of Marx's own writings, and this tendency was further strengthened by the later scientism of Engels.[12] Lenin, of course, was himself profoundly interested in not only the philosophy but also the application of modern science as well.[13] As a measure of his commitment to the scientific ethos, Lenin, in addition to his accommodationist policy towards a largely hostile community of scientists, encouraged a number of individuals and agencies in the early 1920s to work on what might now be called the 'science of science'. Experiments on the 'scientific management' of laboratories (Taylorism), psychological studies of creativity, eugenic research into the genealogies of leading scientists and historical inquiries into the social conditions which fostered or hindered the advancement of science were all vigorously pursued up to and including the first years of the "great break". In fact, the industrial initiatives of the early Stalinist period threw up yet another problem for the theoreticians of science, namely the integration of academic and technological enterprises. The zenith of self-consciousness about scientific development was reached in April of 1931 at the first (and last) All-Union Conference on the Planning of Science. Although the Soviets were not to follow up on any of these initiatives until quite recently, they had already leapt ahead of the world in two vital areas: "(1) recognition of science as a natural resource and the collection of data concerning it, and (2) the posing of legitimate questions about the ways in which a government can aid the development of science."[14]

The practical consequence of such interest was an exponential increase in the financial and human resources available for scientific research after 1928. The following table outlines the enormous expansion which took place in a few selected fields.

The Expansion of Soviet Science, 1928-1934

Activity	1928	1932-33
University Enrollment	159,757	469,215 (1933)
Degree-granting Institutions ...	120	168 (1933)
Post-graduate Scientists	1,548	16,500 (1932)
	Budget (Millions of Roubles)	
	1928	1933-34
Academy of Sciences	3	25 (1934)
Expenditure on Geology ...	10½	140 (1933)

Sources: Alexander G. Korol, *Soviet Research and Development: Its Organization, Personnel and Funds* (Cambridge, Mass., 1965); and A. Pinkevich, *Science and Education in the USSR* (New York, 1935).

On the other hand, the political correlative of increasing state support was a greater degree of state influence over academic activities. Indeed, in a society determined to abolish the hierarchical division of labour, this distinction between science and politics or between "scientific workers" and the proletariat was bound to be viewed as both artificial and pernicious. Yet it would be anachronistic to interpret the situation of Russian scientists during this period in terms of Lysenko's crackdown on genetics during the late 1940s. While acknowledging the absence of detailed studies on lines of basic research extended during the Thirties, it seems likely that, apart from geneticists, most investigators enjoyed a considerable amount of operational freedom.[15] The chief difficulty resided in the persistent threat of sudden and largely unforeseen interventions into entire scientific fields.

It was in this atmosphere of unprecedented growth, heightened self-consciousness and uncertainty about the future that the decision was made to send a delegation to the Second International Congress of the History of Science and Technology. Why this meeting was selected to propagandize the achievements of science under Marxism is not at all clear. Although at least one of the delegates (Zavadovsky) had been scheduled to go to the London meeting for some time,[16] it seems that the commitment to turn up in force was not made until the last minute. Indeed so hasty were the arrangements that, on the plane journey to London, the leader of the Soviet contingent (Nicholai Bukharin) discovered that he had left the text of his speech behind in Moscow![17]

If we are ignorant about the origins of the Russian expedition, we at least know something about the biographies of each of the eight delegates. Bukharin was clearly the most imposing figure within the delegation. A trusted associate of Lenin in the early days of the Revolution, he was expelled by Stalin from the Politburo in 1929 as leader of the "Right deviationists". Despite this political setback (from which he never recovered), Bukharin was able to perform useful services as Director of the Industrial Research Department of the Supreme Economic Council.[18] He was also instrumental in bringing about the reform of the Academy of Sciences, of which he became a leading member after 1929. As President of both the Academy's Commission on the History of Knowledge, and the All-Union Planning Conference, Bukharin was drawn towards the most general theoretical problems posed by the Marxist theory of scientific development.[19] After 1931, though, he was unable to sustain his activities in this area. He was executed in 1938, following one of the most famous purge trials of the Stalinist era.[20]

Four other members of the delegation—Boris Hessen (Gessen in the U.S.S.R.), Ernst Kol'man, B. M. Zavadovsky and V. F. Mitkevich[21]—were primarily concerned with the philosophy of science. With the exception of Mitkevich, they were all Deborinite philosophers prior to the 'great break'. Hessen, whose essay on Newton was to create such a sensation at the 1931 Congress, was best known in Russia as a champion of relativity theory. Originally trained in physics by A. F. Joffe (Ioffe in the U.S.S.R.), Hessen began his career in the Department of the History and Philosophy of the Natural Sciences at Moscow University. His rapid rise in official scientific circles as an approved interpreter of contemporary science was guaranteed by the Deborinite victory over the Mechanists. Despite sporadic challenges from Party philosophers about the Marxist status of Einstein's formulae, Hessen managed to defend himself successfully until 1934. Thereafter, he simply vanished and is generally presumed to have died during one of the purges of the mid-Thirties. Kol'man and Zavadovsky, on the other hand, seem to have been far more adept than Hessen in their ability to abandon positions eventually condemned by the Central Committee. Thus Zavadovsky, while arguing as a 'Morganist' in 1926 that " 'the facts that I have been studying these past years . . . dictate to me the necessity, in the problem of the inheritance of acquired characteristics, of abandoning the views of Darwin and Engels, . . . and Marx,' "[22] wound up a decade later as an early supporter of the Lamarckian arguments of Lysenko and Prezent.[23] The case of Kol'man the mathematician is rather more complicated. A Czech national, he found himself interned in a Russian P.O.W. camp at the end of the First World War. After his release, Kol'man secured a place at Moscow University where he tentatively supported Deborinite physical scientists. His statements on relativity, however, became increasingly oblique during the 'great break'. A few years later Kol'man managed simultaneously to be a partisan of Lysenko[24] and a spokesman for the 'liberal' defence of Einsteinian physics.[25] During the early Fifties he emerged yet again, this time as an advocate of cybernetics, and, at last report, Kol'man appears now to command a schizophrenic reputation as a liberal in Russia and a rigid ideologue in his native Czechoslovakia.[26] In contrast to Kol'man's acrobatics, V. F. Mitkevich appeared straightforward—a highly successful electrical engineer who became a hero to young scientific 'Bolshevizers' in 1931 for his opposition to the 'formalist' tendencies of contemporary physics.[27] To counter such trends Mitkevich called for a return to the visual models favoured by Faraday and other nineteenth-century theorists.

The remaining three delegates were administrators or working scientists who took little part in the philosophical controversies of this period. Modest I. Rubinstein, was an economist, attached to the Communist Academy during the early Thirties. While many of his interests in technical policy overlapped those of Bukharin, Rubinstein somehow survived the political turbulence of the late 1930s. Since 1945 he has devoted himself to historiographical studies in general and the history of science and technology in particular.[28] Abram Feodorovitch Ioffe was the doyen of Soviet physicists until his death in 1960 at the age of eighty. During the inter-war period he oversaw the training of an entire generation of researchers in the physical sciences.[29] A sincere patriot, Ioffe joined the Communist Party in 1942, but his interest in Marxism was essentially limited to the achievement of prosperity and social justice through the creation of an efficient and prosperous economy.[30] Finally we come to the last member of the Russian contingent, N. I. Vavilov. Even before Medvedev's moving account of the first phase of the Lysenkoist debate, it had been clear that Vavilov's courageous defence of genetics, which led to his tragic death in 1942, constituted one of the greatest single acts of martyrdom in the history of science.[31] Yet, in 1931, Vavilov was a self-confident and successful plant geneticist. As President of the Lenin Academy of Agricultural Sciences, he already enjoyed a mammoth budget and a world-wide reputation for his research on the historical origins of agriculture, not to mention his unparalleled collection of grain seedlings.

In retrospect, we may say that the members of the Soviet delegation were well qualified to speak about the history and direction of science since the Revolution. Unlike most of the historians, philosophers and scientists whom they were eventually to confront in London, the Russians offered their scholarship as a contribution to a programme of socialist reconstruction which relied heavily on the work of natural scientists. The schism between Marxists and non-Marxists at the Congress would have been even greater, in fact, had the Soviet Union been represented by the more ardent of the Bolshevizers, who flatly rejected many of the scientific theories then prevalent among Western (and Soviet) scientists. As it happened, neither Bukharin nor the Deborinites nor the administrators and other scientists within the Russian contingent were inclined in this period to label widely held, contemporary paradigms as 'bourgeois' or 'undialectical'. What they wished to communicate above all else was the intellectual vitality, self-awareness, social usefulness and sheer prosperity of science in a socialist society.

Turning to Britain, we find that there were by 1931 several scientists, including J. D. Bernal, J. B. S. Haldane, Lancelot Hogben, Hyman Levy and Joseph Needham, who were prepared to be moved by what the Russians had to say.[32] During the preceding decade, they had all shown a growing commitment to socialism. These researchers had also been exposed in varying degrees to Marxism, and had even developed an interest in the historical and philosophical problems which concerned many Soviet scientists. For the most part, however, the participation of this group in local and national politics had been both limited in extent and rigorously separated from their professional lives.[33] Bernal was unique at the time in his thoroughgoing Marxism, as well as in his Communist Party membership. Just as unusual was Levy's long service (1924-1930) as Chairman of the Labour Party's Science Advisory Committee. On the question of historical studies,

Bernal and Hogben were already impressive amateurs,[34] but it was Joseph Needham who rapidly achieved a high reputation for his erudite studies on the history of embryology.[35] Hogben and Needham had also gained a certain amount of notoriety in the late Twenties for their opposition to what they regarded as the neo-vitalistic philosophies of senior biologists like J. S. Haldane and E. S. Russell.[36] On the strength of their earlier work in the history and philosophy of science, Hogben and Needham were asked to participate at the 1931 Congress in the dual roles of participants in the session on 'The Historical and Contemporary Inter-Relationship of the Physical and Biological Sciences' and organizers of the London meeting. Needham, who was a friend of Charles Singer, the Congress's presiding officer, sat on the Executive Committee; Hogben served on the Council.

In the light of their collective biography, one is not surprised to learn that the social and political crises which accompanied the onset of the Great Depression intensified the socialist inclinations of this group of scientists. It was not just that, in 1931, a faltering economy was resulting in misery for unemployed millions and the downfall of the second Labour government,[37] but also that scientific research was itself beginning to suffer from financial restrictions. The slow increase of government funds for academic and industrial research which had characterised the Twenties was stopped and then reversed. When unemployment became a reality for highly trained specialists (especially chemists),[38] younger scientists were able to recognise, in the words of Lancelot Hogben, "which side of their bread had no butter."[39] The more prominent of the socialist researchers, however, were drawn to the larger aspects of the political situation. Unlike their first years in the laboratory, detachment from society now seemed an intolerable pose: "I tried to keep to my own field," Joseph Needham would later write, "but politics would keep breaking in."[40]

Having engaged themselves in the incipient political upheaval of the Depression era, these 'scientific workers' found themselves without analytical tools to understand the past and to provide guidance for the future.[41] Their earlier sanguine attitude towards the economy of capitalist Britain perhaps lay at the root of their later difficulties.[42] Obviously the most fully developed working alternative to the capitalist society in which they worked was post-Revolutionary Russia. But prior to 1931, their interest in Soviet events was ill-defined and sporadic. Bernal's comments in the late Twenties about the possibility of constructing a fully scientific régime in the U.S.S.R., while suggestive, were uninformed.[43] Only Haldane, in fact, had met the Russians on their own territory (in

1928), and, although his subsequent reports were largely favourable, he also expressed a certain anxiety about the future of genetics in a Marxist society.[44] Thus the information of the English Marxists about the state of Soviet science was not only fragmentary but pre-dated the critical events of the 'great break' as well. It was only in the midst of a later social crisis that such figures as Bernal, Hogben and Levy were stimulated to find out what was going on in Russia. There to supply the answers—in London—were the Soviet delegates to the Second International Congress of the History of Science and Technology.

The date and site of the Congress were chosen on an *ad hoc* basis by the International Academy of the History of Sciences (founded in 1928).[45] Quite naturally, the timing of this international gathering of scholars was determined without regard to the coming but unforseeable economic catastrophe. The consequences of meeting at the height of the Depression were obvious enough for British socialists. Yet an even more important point remains to be made. Between the early Twenties and the late Sixties, the Soviets were never so well briefed or so highly motivated to discuss the implications of 'Marxist science' as they were in the spring of 1931. A delay of one or two years on either side, quite apart from the effect of shifting the Congress into periods of prosperity or recovery, could well have resulted in one of two possibilities: the failure of the Russians to turn up at all, or the appearance of a delegation significantly different in terms of personnel and even intellectual orientation from the one selected in 1931.[46] As for the site, the choice of London rested, in part at least, on the fact that the British Association for the Advancement of Science would be meeting for the first time in the Kingdom's capital to celebrate its centenary year. Hence, there was never any question that, if the Russians were to make any sort of impression at all, it would be in England.

Russians in London

From their arrival in London by aeroplane—something of a novelty at the time—to the large reception given in their honour at the Soviet Embassy, the Russian delegates were quite easily the most conspicuous group at this international congress. J. G. Crowther, science correspondent for the *Manchester Guardian*, informed his readers that "the Russian government is taking an important part in the International Congress of the History of Science and Technology."[47] Whatever prominence the delegation enjoyed, Congress officials could not be persuaded either to expand the time allotted for speakers in the scheduled sessions or to cut back on the allotted number of social functions.[48] A compromise was reached in the creation of a special meeting on Saturday.

From the beginning, the Soviets injected themselves into almost all of the Congress's formal discussions. When on June 30th, for example, the historian G. N. Clark and the physiologist A. V. Hill suggested that the subject matter of history should be broadened to include the record of intellectual advances, *The Times*' correspondent noted that "some interest was caused . . . by the request of five members of the Soviet delegation . . . that they might be allowed to contribute a Marxian view to the symposium."[49] The upshot of their comments was that the recommendations of Clark and Hill would only result in new forms of hero worship, i.e., Newton and Darwin over Marlborough and Lincoln.[50] What was required instead was a move away from individualistic or 'bourgeois' philosophies of history, towards the Marxist approach, which emphasised the way in which great men have been moulded by the social and economic forces of their time.

Two days later, in the session devoted to the philosophy of biology, the Russian speaker, B. M. Zavadovsky, allied himself with Lancelot Hogben and Joseph Needham in a joint attack on the positions of J. S. Haldane and E. S. Russell.[51] After Hogben had wondered aloud whether the resurgence of antipathy towards mechanism and materialism might not be due to popular fear in the West of Bolshevism and social unrest, Zavadovsky went on to assert that "these tendencies characterize the general disillusionment of bourgeois society in the possibilities of material culture".[52] The Soviet biologist added that, for the dialectical materialist, the dilemma posed by vitalism-mechanism was a false one. To admit that the biological level was qualitatively distinct from that of the physical merely implied the need for novel experimental tools and a subtler version of the materialist viewpoint. The overall framework, Zavadovsky contended, had already been laid out "in the classical works of . . . Marx and Engels, and, in our own times, in the profound works of Lenin."[53]

After several days of philosophical interventions, the Russians concluded the Congress with a more detailed presentation of the Marxist approach to the history and planning of science. Since this volume contains the addresses delivered at that meeting, it may be useful to focus attention upon the highly controversial piece by Hessen.[54] His arguments about Newton require close analysis not only because of their later impact, but because they also illustrate the utilitarian rationale which underlay so much of Soviet scholarship in this period.

In a certain sense, the *Principia* is merely a peg on which Hessen is able to hang a highly compressed history of the physical sciences from the seventeenth to the nineteenth centuries. For Hessen, the chief feature of the Newtonian era is the rise of capitalism which created, first, new demands for technology and, second, deep political and religious divisions within English society. After listing a series of technical problems related to ballistics, hydro-statics, magnetism, optics and mechanics, Hessen asserts that "the 'earthy core' of the *Principia* consists of just those technical problems which we have analysed above, and which fundamentally determined the themes of physical research of the period."[55] By this statement, Hessen does not mean that Newton's *analytical* methods were directly related to economic factors. Here we are enjoined to take into account the superstructure of philosophical theories and religious beliefs which guided Newton along certain lines of thought.[56] Hessen thus dissects the *Principia* into its idealist, mechanist and materialist components, in order to suggest that Newton's great work was the philosophical equivalent of the social and political compromises of the late seventeenth century. He then concludes that what is soundest in the writings of Newton can be correlated with the technical rather than the ideological demands of the time.[57] The moral which Hessen draws from all this is that science cannot advance in a society which restricts technological advancement: "Science develops out of production, and those social forms which become fetters upon productive forces likewise become fetters upon science."[58] At the close of his speech, Hessen confidently draws a parallel between the English and Russian Revolutions, when he remarks that, "as in all epochs, in reconstructing social relationships we are reconstructing science."[59]

When Hessen finished his talk, there was a long and awkward silence which was finally punctuated by the twenty-year-old Cambridge mathematics student, David Guest. With the encouragement of Hyman Levy, Guest offered support for Hessen's thesis by applying it to the 'contradictions' inherent in modern physical theory.[60] Only two English historians of science, Whetham and Wolf, voiced any opposition. This prompted J. D. Bernal to remark shortly afterwards that the Russians' papers probably had little effect on most of the audience. The appeal to Marxism, instead of impressing their listeners, probably "disposed them not to listen to the arguments which followed, with the feeling that anything so ungentlemanly and doctrinaire had best be politely ignored."[61]

The performance of the Soviet delegation was not overlooked, however, by "*The Times* of the scientific community," the journal *Nature*.[62] In a review of the Congress, Thomas Greenwood described historical materialism as a "communistic explanation of scientific development, in which the integrative work of the masses is exalted at the expense of the glorification of genius." He further maintained that "the attitude of the Soviet delegates can scarcely explain any history, however stimulating their message and their endeavours to put it into practice in their own educational institutions."[63] Three weeks later, in a review of *Science at the Cross Roads*, the historian F. S. Marvin was rather more generous. He agreed that knowledge was, in part, a social product, but he doubted whether such an insight could account for all aspects of the history of science. Marvin then went on to condemn the concept of "bourgeois" science: "The laws of nature are the same for all of us".[64] In bringing his remarks to a close, he expressed anxiety over the effect which dialectical materialism might have on the direction of research in the Soviet Union. It was evident that, while the appearance of the Russians at the Congress did merit *Nature*'s attention, it did not mark a turning point in the journal's views about science and society.[65]

But its impact upon the British socialist scientists discussed above was profound. Scientifically and politically, they found themselves in substantial agreement with the Soviets. Zavadovsky's paper, for instance, greatly impressed Needham, because its conclusions—so similar to his own—had apparently been derived from the axioms of dialectical materialism.[66] The contribution of Hessen made an even deeper and wider impression. As Hogben has since recalled, it "reinforced my interest" in historical materialism "as an intellectual tool for expository use."[67] Hyman Levy suddenly found most works on the history of science inadequate, because they did not "give an account at the same time of man's social and economic background."[68] For J. G. Crowther, Hessen's essay fundamentally determined the direction of all his later work in the history of science.[69] Bernal's opinion has already been cited (ref. 3). But perhaps the most interesting convert to the Marxist viewpoint was Joseph Needham. When revising the first volume of his massive *Chemical Embryology* for its republication as *A History of Embryology*, Needham argued that scientific advances must not be dissociated from technical needs and processes and then added that "further historical research will enable us to do for the great embryologists what has been so well done by Hessen for Isaac Newton".[70]

Yet, however fascinated these men were with the new vistas afforded by historical and dialectical materialism, it was the political message of the Soviets which most immediately concerned them. Bernal, Crowther, Hogben, Levy and Needham had ample opportunity at the Congress itself to discuss the state of science in Russia with the Soviet delegates, particularly Bukharin and Hessen.[71] As the Russians flew back home, they left Bernal and his associates with the following dilemma: "Is it better to be intellectually free but socially totally ineffective, or to become a component part of a system where knowledge and action are joined for one common social purpose?"[72]

The Aftermath of the Congress

In the short space which remains, it is instructive to sketch some of the developments in the history, philosophy and planning of science which owe something to the publication of *Science at the Cross Roads*.[73]

The popularity of the Marxist approach to the history of science (in America and Europe) has varied considerably since the appearance of Hessen's paper. Since the nadir of this historiography in the West coincided with the Cold War era, we cannot overlook the subtle (and occasionally crude) forms of political intimidation which operated *within* the scholarly world to the detriment of a developing Marxism.[74] The revelations arising out of the Lysenko affair reinforced the tide of anti-communism. But two other factors must be taken into account. The first is that, even in England, the Thirties' Marxists who were interested in the history of science always worked from a weak institutional base. Apart from the absence of journals which might have sustained this tradition,[75] none of the key figures—Bernal, Gordon Childe, Crowther, Benjamin Farrington, Christopher Hill, Eric Hobsbawm, Hogben, Sam Lilley and Needham—was himself in a position to train a new generation of professional historians of science. Second, after 1945 the history of science emerged as a distinct academic discipline under the guidance of scholars supremely conscious of the Marxists' neglect of science as a body of ideas.[76] Only since the early Sixties have systematic attempts also been made to consider science and technology as agents *and* products of social change themselves.[77]

With the exception of the work of David Bohm and the successors of Gaston Bachelard, there has been no comparable resurgence of interest in a philosophy of science based on dialectical materialism. Even at Marxism's high tide in England, the dialectical approach was never pursued with much vigour beyond the level of very general discussions.[78] The most notable work was that of Bernal and Haldane on the origin of life.[79] It is true that between 1945 and 1950 some philosophers and writers advanced some provocative theses in this area,[80] but they were quickly overshadowed by the furore over Lysenkoism.

As for theories about the planning of science, one man, J. D. Bernal, has until very recently dominated the field. His classic work on *The Social Function of Science* successfully predicted many of the post-war period's most central features, i.e., the expansion of scientific activities in national life, the co-ordination of academic research centres and the evolution of national science policies.[81] Although it is questionable whether his work had much direct influence on government operations in Britain[82] and the United States, it does seem to have made some impact in France[83] and in the Soviet Union. This fact, of course, brings us full circle. When we realise that an important part of Bernal's thinking on planned science derived from Bukharin (*persona non grata* to this day in the U.S.S.R.), we cannot miss the irony of this intellectual debt here repaid. The high esteem in which Bernal is held in Russia has thus helped scholars in that country to return circuitously to the work of their fellow-countrymen in the 1920s.[84]

Forty years after the publication of this book, we are beginning to sense that, once again, science is finding itself at the crossroads. Viewed as the instruments of those in power, it is inevitable that, in a time of bitter conflict over what the structure and goals of society ought to be, various scientific communities will be subjected to a barrage of criticism. One of the most interesting features of the recent and continuing debate on the social rôle of modern science has been the anti-technocratic stance of much of the New Left. Viewed from the standpoint of Marxism's long-standing commitment to the scientific world-view, it may not be too far fetched to suggest that we are now witnessing a fundamental divide in the history of radical thought. Before this divide is crossed, however, it will be necessary to come to terms with the older view that technocratic means and socialist ends are not incompatible. Undoubtedly one of the key documents in this re-evaluation will be *Science at the Cross Roads.*

NOTES

I am grateful to Drs. R. M. MacLeod and R. M. Young for their comments on the first draft of this essay.

1 Hessen's essay on "The Social and Economic Roots of Newton's 'Principia'," was later reproduced as a pamphlet (Sydney, 1946).

2 On Soviet science in this period, I have relied heavily on two works: David Joravsky, *Soviet Marxism and Natural Science, 1917—1932* (London, 1961); and Loren R. Graham, *The Soviet Academy of Sciences and the Communist Party, 1927—1932* (Princeton, 1967).

3 J. D. Bernal, *The Social Function of Science* (London, 1939; reprinted Cambridge, Mass., 1967), p. 406.

4 Hyman Levy, *Modern Science* (London, 1939), p. 97.
5 John R. Baker and A. G. Tansley, "The Course of the Controversy on Freedom in Science," *Nature, 158* (October 26, 1946), 574. Later commentators have supported—in far more neutral language—Baker and Tansley. See Neal Wood, *Communism and British Intellectuals* (London, 1959), esp. pp. 123-25. Cf. Michael D. King, "Science and the Professional Dilemma," in Julius Gould (ed.), *Penguin Social Sciences Survey 1968* (Harmondsworth, 1968), p. 57.
6 Baker and Tansley, *loc. cit.*, 574.
7 Levy, *Modern Science*, p. 97.
8 This conclusion is based on research material in my doctoral dissertation, "The Visible College: A Study of Left-wing Scientists in Britain, 1918—1939" (Harvard University, in preparation).
9 On the Deborinite-Mechanist controversy, see Joravsky, *Soviet Marxism*, pp. 93-214.
10 Graham, *Soviet Academy*, pp. 120-53.
11 David Holloway raises a number of interesting points on this subject in his "Scientific Truth and Political Authority in the Soviet Union," *Government and Opposition, 5* (Summer 1970), 345-67.
12 The chief texts are, of course, Engels's *Anti-Dühring* and his *Dialectics of Nature.* For a Marxist critique of the scientistic tendencies in these works, see George Lichtheim's *Marxism: An Historical and Critical Study* (New York, 2nd ed., 1965), esp. pp. 234-58.
13 See V. I. Lenin, *Materialism and Empirio-criticism* (Moscow, 1947). On Lenin's interest in science, there is I. I. Artobolevsky and A. A. Chekanov, "V. I. Lenin and Science," *Scientific World, 14* (1970), 4-6.
14 Graham, *Soviet Academy*, p. 62.
15 For a general account of Soviet science in the late Thirties and early Forties, see Eric Ashby, *Scientist in Russia* (Harmondsworth, 1947). On the early crisis in genetics, see Z. A. Medvedev, *The Rise and Fall of T. D. Lysenko* (New York, 1969). Unfortunately, I have been unable to see David Joravsky's recent *The Lysenko Affair* (Cambridge, Mass., 1970). The development of an important sub-discipline within Russian genetics is well treated by Mark B. Adams in his "The Founding of Population Genetics: Contributions of the Chetverikov School, 1924—1934," *Journal of the History of Biology, 1* (1968), 23-39.
16 Zavadovsky is listed as a speaker in the original prospectus of the Congress.
17 J. G. Crowther, *Fifty Years with Science* (London, 1970), p. 77. This source contains the best description (pp. 76-80) of the events surrounding the arrival of the Russians at the Congress.
18 It was not unusual for highly talented "Right deviationists" to hold highly responsible positions in the State bureaucracy during the early 1930s, especially if they enjoyed the confidence of their officially approved superiors. I am grateful to Professor R. W. Davies for this information.
19 In addition to his address in this volume, see Bukharin's article in N. I. Bukharin and others, *Marxism and Modern Thought* (London, 1935), pp. 1-90. Note as well the comments on Bukharin's openness to contemporary scientific theories in Joravsky, *Soviet Marxism*, pp. 99-104.
20 See G. Katkov, *The Trial of Bukharin* (London, 1970).
21 Mitkevich is listed in *Science at the Crossroads* (Kniga, 1931) as W. Th. Mitkewich. Nevertheless, there is no doubt that V. F. Mitkevich and

W. Th. Mitkewich are one and the same person: Joravsky, *Soviet Marxism*, pp. 289-95.

22 B. M. Zavadovsky, "Darvinism," *Vestnik kommunisticheskoi akademii*, 1926, Kn. 14, p. 273; as quoted in Joravsky, *Soviet Marxism*, p. 218.

23 Medvedev, *Rise and Fall of Lysenko*, p. 21. Zavadovsky's brother, M. M. Zavadovsky, was an anti-Lysenkoist.

24 *Ibid.*, p. 83.

25 See Loren R. Graham, "Quantum Mechanics and Dialectical Materialism," *Slavic Review, 25* (1966), 408.

26 *Idem.* On Kol'man's role as a spokesman for the cyberneticians, see David Holloway's forthcoming book on the introduction and development of cybernetics in the Soviet Union.

27 Joravsky, *Soviet Marxism*, pp. 289-95.

28 Note, for example, his review of the Russian translation of J. D. Bernal's *Science in History* (London, 1954) in *Kommunist*, 1957, *14* (October), 110-17. I am grateful to David Holloway for this reference, as well as to Robert A. Lewis for information about Rubinstein's biography.

29 One of his most famous students was Peter Kapitsa: see Albert Parry (ed.), *Peter Kapitsa on Life and Science* (New York, 1968), esp. pp. 75-121.

30 See Joravsky, *Soviet Marxism*, pp. 287-90. Cf. Graham, *Soviet Academy*, p. 26.

31 A biography of Vavilov has recently appeared in the Soviet Union: see S. Ye. Reznik, *Nikolai Vavilov* (Moscow, 1968). Vavilov's brother, Sergei, later became President of the Soviet Academy of Sciences and was coerced into the Lysenkoist camp in the late Forties.

32 These are the principal characters in my "Visible College". As far as I know, Haldane was not in London during the 1931 Congress.

33 One scientific issue which did provoke their political consciousness was eugenics. See, for example, Lancelot Hogben, *The Nature of Living Matter* (London, 1930), esp. pp. 193-214.

34 Thus Allen Hutt once recalled that the substance of much of the historical sections in *The Social Function of Science* had been the subject of undergraduate conversations between Bernal and himself. See Hutt's "Science and Society," *Labour Monthly, 21* (June 1939), 319.

35 See the first volume of Needham's *Chemical Embryology* (Cambridge, 1931).

36 An account of this controversy is contained in the fourth chapter of "The Visible College".

37 See Robert Skidelsky, *Politicians and the Slump: the Labour Government*, 1929-1931 (London, 1967). Cf. Eric Hobsbawm, *Industry and Empire* (Harmondsworth, 1969), pp. 207-24.

38 See Dennis Chapman, "The Profession of Chemistry during the Depression," *Scientific Worker*, (Autumn 1939), 74-81.

39 Lancelot Hogben, *Science in Authority* (London, 1963), p. 118.

40 Joseph Needham, *Time: the Refreshing River* (London, 1943), p. 11.

41 The problems faced by the scientists in this regard reflected a general intellectual weakness in the British socialist tradition, particularly on the sociological side. See Philip Abrams, *The Origins of British Sociology, 1834-1914* (Chicago, 1968), pp. 3-153; and Perry Anderson, "Components of the National Culture," *New Left Review, 53* (July/August, 1968), 3-57.

42 Note, for instance, the comments of J. B. S. Haldane in his *Daedalus, or Science and the Future* (London, 1923), esp. p. 6. Cf. J. D. Bernal, *The World, the Flesh and the Devil* (London, 1929), pp. 70-71. Indiana University Press has recently reprinted this work (Bloomington, 1969).

43 Bernal, *World, Flesh, Devil,* pp. 93-94.

44 See J. B. S. Haldane, *The Inequality of Man, and Other Essays* (London 1932), pp. 126-36.

45 The first Congress organised by the Academy met in Paris in 1929. On the foundation and early history of the Academy, see the following works by George Sarton: "Acta Atque Agenda," *Actes due VIe Congrès International d'Histoire des Sciences* (Paris, 1951), esp. pp. 62-66; and *The Study of the History of Science* (Cambridge, Mass., 1936), pp. 66-70.

46 Before 1934 there was some contact between left-wing scientists in France and a few members of the Soviet delegation of 1931. See David Caute, *Communism and the French Intellectuals, 1914—1960* (London, 1964), esp. pp. 300-17. Otherwise, it became extremely difficult after the early Thirties for Russian scholars to travel outside their own country. One of the earliest indications of such restriction was the cancellation of the 1937 International Congress of Genetics which was to have been held in Moscow. For contrasting but equally concerned British reactions to the isolation of Soviet science, see: B. P. Uvarov, "Genetics and Plant Breeding in the U.S.S.R.," *Nature, 140* (21 August, 1937), 297; and Joseph Needham and Jane Sykes Davies (eds.), *Science in Soviet Russia* (London, 1942), pp. 27-28.

47 "A Congress of Scientists," *Manchester Guardian,* June 30, 1931, p. 8. I am grateful to Dr. Arnold Pacey for his last-minute research on Crowther's articles in the *Guardian* during the week of the Congress.

48 Thus the Congress was recessed from Tuesday afternoon, June 30th, to Thursday morning, July 2nd, to allow delegates to travel to Cambridge for a lunch and some exhibitions organised by Trinity College.

49 *The Times,* July 1, 1931, p. 16.

50 This information comes from J. D. Bernal's article on the Congress for *The Spectator* (July 11, 1931) as reprinted in his *The Freedom of Necessity* (London, 1949), p. 335.

51 A set of the cyclostyled sheets of the addresses delivered at this session can be found in the private papers of Dr. Joseph Needham.

52 B. M. Zavadovsky, "The 'Physical' and the 'Biological' in the Process of Organic Evolution," fourth essay in *Science at the Crossroads,* p. 2.

53 *Ibid.,* p. 12.

54 The essays by Bukharin, Ioffe and Rubinstein deal with the theory and practice of science planned to meet agricultural and industrial needs. Kol'man and Mitkevich, in their very different ways, criticise the 'formalist' tendencies of modern physics. As for the non-Marxist Vavilov, reference to Marxism is kept to a minimum in his description of Soviet research on plant genetics.

55 Boris M. Hessen, "The Social and Economic Roots of Newton's 'Principia'," ninth essay in *Science at the Crossroads,* p. 21.

56 *Ibid.,* pp. 27 ff.

57 *Ibid.,* p. 40.

58 *Ibid.,* p. 60

59 *Ibid.,* p. 62.

60 See Hyman Levy, "The Mathematician in the Struggle," in Carmel

Haden Guest (ed.), *David Guest: A Scientist Fights for Freedom* (London, 1939), pp. 152-53. Cf. J. G. Crowther, *The Social Relations of Science* (New York, 1941), pp. 616-17.

61 Bernal, *Freedom of Necessity*, p. 338.

62 On *Nature's* political orientation during the interwar period, see Paul Gary Werskey, "*Nature* and Politics between the Wars," *Nature, 224* (November 1, 1969), 462-72. Cf. W. H. G. Armytage, *Sir Richard Gregory, His Life and Work* (London, 1957).

63 Thomas Greenwood, "The Third (sic) International Congress of the History of Science and Technology," *Nature, 128* (July 11, 1931), 78.

64 F. S. Marvin, "Soviet Science," *Nature, 128* (August 1, 1931), 170-71.

65 *Nature's* pronounced hostility towards Soviet science in the late Thirties is documented in P. G. Werskey, "British Scientists and 'Outsider' Politics, 1931—1945," *Science Studies, 1* (January, 1971), 76-77.

66 Needham, *Time: the Refreshing River*, p. 244.

67 Lancelot Hogben, personal communication to the author (July 26, 1968), in which he also remarks about his long-standing antipathy to *dialectical* materialism.

68 Hyman Levy, *Science in Perspective: An Essay Introductory to Twenty-four Talks* (London, 1931), p. 47. These talks were broadcast over the B.B.C. and were published in Mary Adams (ed.), *Science in the Changing World* (London, 1933).

69 J. G. Crowther, *British Scientists of the Nineteenth Century* (London, 1935), p. ix.

70 Joseph Needham, *A History of Embryology* (Cambridge, 1934), p. xvi.

71 Bukharin managed to visit the homes of both Hogben and Levy while he was in London. I learned this in separate conversations with the two British scientists.

72 Bernal, *Freedom of Necessity*, p. 339.

73 On the growth of political radicalism among British scientists during the Thirties, see "The Visible College," and P. G. Werskey "Radical Cambridge: Left-wing Scientists in the 1930s," (typescript, Science Studies Unit, University of Edinburgh) which will appear in a forthcoming volume of essays edited by Dr. R. M. MacLeod on British science in the interwar period.

74 See, for example, Henry Guerlac, "Some Historical Assumptions of the History of Science," in A. C. Crombie, *Scientific Change* (London, 1963), esp. pp. 198-201. For recent criticism of the historian's neglect of these subjects, Cf. Arnold Thackray, "Science: Has Its Present Past a Future?" in Roger H. Stuewer (ed.), *Historical and Philosophical Perspectives of Science* (Minneapolis, 1970), pp. 112-127.

75 *Labour Monthly* and the *Modern Quarterly*, which, at different points in the 1930s, devoted some attention to the history of science, were drawn away from this area almost completely after 1945.

76 In England, the first sustained criticism of the Marxist historiography of science—laced with a number of striking concessions—was G. N. Clark's *Science and Social Welfare in the Age of Newton* (Oxford, 1937). The book which seems to have established a 'paradigm' for British historians of science in the early post-war period was Herbert Butterfield's *The Origins of Modern Science, 1300—1800* (London, 1949). Butterfield's protegé, A. Rupert Hall, then went on to produce *Ballistics in the Seventeenth Century* (Cambridge, 1952), in which he, as it were, stood Hessen's essay on its head. In attempting to account

for changing patterns in the direction of the history of science discipline in England, it would be fascinating to study the evolution of the Cambridge Committee on the History of Science, which was chaired by Joseph Needham before the Second World War but controlled by Butterfield thereafter. On the origins and early activities of the Cambridge Committee, see Joseph Needham and Walter Pagel (eds.), *Background to Modern Science* (Cambridge, 1938), pp. vii-xii. Such a study could well reveal the futility of drawing rigid distinctions between the academic and the political.

77 See T. S. Kuhn, "The Relations between History and the History of Science," *Daedalus, 100* (Spring 1971), esp. 298-300. The chief works in this area derive from "critical theory": J. Habermas, *Toward a Rational Society* (London, 1970); and Herbert Marcuse, *One-Dimensional Man* (Boston, 1964).

78 Besides the essays of Bernal and Levy in *Aspects of Dialectical Materialism* (London, 1934), see the following: Bernal's *Freedom of Necessity*, pp. 334-428, and his *Marx and Science* (London, 1952); J. B. S. Haldane, *The Marxist Philosophy and the Sciences* (London, 1938); Levy's *Philosophy for a Modern Man* (London, 1938), and his *Modern Science;* and Joseph Needham, *Order and Life* (New Haven, 1935; Cambridge, Mass., 1968). Note as well Christopher Caudwell, *The Crisis in Physics* (London, 1939).

79 See J. D. Bernal, *The Origin of Life* (London, 1967).

80 In particular, Maurice Cornforth, *Science versus Idealism* (London, 1946); and Jack Lindsay, *Marxism and Contemporary Science* (London, 1949).

81 See the OECD series "Reviews of National Science Policy". For Bernal's reactions to these developments, see his essay "After Twenty-five Years," in Maurice Goldsmith and Alan MacKay, *The Science of Science* (Harmondsworth, 1966), pp. 285-309. This article is also appended to the 1967 edition of *The Social Function of Science*, pp. xvii-xxxvi.

82 See Norman J. Vig, *Science and Technology in British Politics* (Oxford, 1968).

83 Bernal's influence on Frédéric Joliot-Curie and French scientific administration is discussed briefly in Pierre Biquard, *Frédéric Joliot-Curie, The Man and His Theories* (New York, 1966), pp. 96-102 and 142.

84 David Holloway has kindly called my attention not only to this point but also to what is possibly the first allusion to have been made in Russia to *Science at the Crossroads* since the early 1930s. See G. M. Dobrov (ed.), *Potensial Nauki* (Kiev, 1969), p. 5, for the following statement (as translated by David Holloway): "substantial influence on the development of science studies was exercised by sociological research carried out in the Soviet Union in those years [the interwar period] into the experience of world science, and set out particularly at the second International Congress on the History of Science."

FOREWORD.

THIS BOOK is a collection of the papers presented to the Second International Congress of the History of Science and Technology by the Delegation of Soviet Scientists. In Soviet Russia absolutely new prospects are opening before science. The planned economy of Socialism, the enormous extent of the constructive activity—in town and village, in the chief centres and in the remotest parts—demand that science should advance at an exceptional pace. The whole world is divided into two economic systems, two systems of social relationship, two types of culture. In the capitalist world the profound economic decline is reflected in the paralysing crisis of scientific thought and philosophy generally. In the Socialist section of the world we observe an entirely new phenomenon: a new conjunction of theory and practice, the collective organisation of scientific research planned on the scale of an enormous country, the ever-increasing penetration of a single method—the method of Dialectical Materialism—into all scientific disciplines. Thus the new type of intellectual culture, which dominates the mental activity of millions of workers, is becoming the greatest force of the present day. The papers here presented to the reader reflect to some extent this great process of social transformation which is taking place in our time. Consequently it is to be hoped that they will be of interest to all who are pondering over the vexed question of the immediate future in the development of human society.

CONTENTS.

THE PROBLEM OF THE ORIGIN OF THE WORLD'S AGRICULTURE IN THE LIGHT OF THE LATEST INVESTIGATIONS.

N. I. VAVILOV, Member of the Academy of Sciences, President of the Lenin Agricultural Academy.

THE WORK OF FARADAY AND MODERN DEVELOPMENTS IN THE APPLICATION OF ELECTRICAL ENERGY.

W. TH. MITKEWICH, Member of the Academy of Sciences.

ELECTRIFICATION AS THE BASIS OF TECHNICAL RECONSTRUCTION IN THE SOVIET UNION.

M. RUBINSTEIN.

THE SOCIAL AND ECONOMIC ROOTS OF NEWTON'S 'PRINCIPIA.'

B. HESSEN, Director of the Moscow Institute of Physics, Member of the Presidium of the State Scientific Council.

THE PRESENT CRISIS IN THE MATHEMATICAL SCIENCES AND GENERAL OUTLINE FOR THEIR RECONSTRUCTION.

E. COLMAN.

SHORT COMMUNICATION ON THE UNPUBLISHED WRITINGS OF KARL MARX DEALING WITH MATHEMATICS, THE NATURAL SCIENCES, TECHNOLOGY, AND THE HISTORY OF THESE SUBJECTS.

E. COLMAN.

THEORY AND PRACTICE FROM THE STANDPOINT OF DIALECTICAL MATERIALISM.

By N. I. BUKHARIN.

THEORY AND PRACTICE FROM THE STANDPOINT OF DIALECTICAL MATERIALISM.

By N. BUKHARIN.

The crisis of present-day capitalist economy has produced a most profound crisis in the whole of capitalist culture; a crisis in individual branches of science, a crisis in epistemology, a crisis in world outlook, a crisis in world feeling. In such historical circumstances the question of the *interrelations between theory and practice* has also become one of the most acute problems, and, moreover, as a question *both* of theory *and* of practice simultaneously. Therefore we have to examine problems from various aspects: (*a*) as a problem of epistemology, (*b*) as a problem of sociology, (*c*) as a problem of history, (*d*) as a problem of modern culture. Lastly, it is interesting (*e*) to verify the corresponding theoretical conceptions from the gigantic experience of the revolution, and (*f*) to give a certain forecast.

1.—*The epistemological importance of the problem.*

The crisis in modern physics—and equally in the whole of natural science, plus the so-called mental sciences (Geisteswissenschaften)—has raised as an urgent problem, and with renewed violence, the fundamental questions of philosophy: the question of the *objective reality of the external world*, independent of the subject perceiving it, and the question of its *cognisability* (or, alternatively, non-cognisability). Nearly all the schools of philosophy, from theologising metaphysics to the Avenarian-Machist philosophy of "pure description" and renovated "pragmatism," with the exception of dialectical materialism (Marxism), start from the thesis, considered irrefutable, that "I" have been "given" only "my" own "sensations."[1]

[1] Cf. *Ernst Mach*: "Analyse der Empfindungen," and his "Erkenntnis und Irrtum; *K. Pearson*: "The Grammar of Science, Lond. 1900. *H. Bergson*: "L'évolution créatrice," Paris, F. Alcan, 1907. *W. James*: "Pragmatism," N. York, 1908, and his "The Varieties of Religious Experience," Lond. 1909. *H. Vaihinger*: "Die Philosophie des Als Ob," Berlin, 1911. *H. Poincaré*: "La Science et l'Hypothèse," Paris, E. Flammarion, 1908. In the same circle of ideas there moves the "logistics" of *B. Russell.* The latest literature on this subject includes the work of *Ph. Frank, M. Schlick, R. Carnap,* et al. Even the almost materialist *Study* takes his stand on the principle quoted: cf. his "Die realistische Weltansicht und die Lehre vom Raume," I. Teil: Das Problem der Aussenwelt. 2. ungearbeitete Aufl, Vieweg & Sohn, 1923.

This statement, the most brilliant exponent of which was Bishop Berkeley,[2] is quite unnecessarily exalted into a new gospel of epistemology. When, for example, *M. Schlick*[3] on this basis builds up a completely "final" ("durchaus endgültige") turning point in philosophy, it sounds quite naive. Even *R. Avenarius*[4] thought it necessary to emphasise all the instability of this initial "axiom." Yet at the present time Berkeley's thesis is strolling up and down all the highways of modern philosophy, and has become rooted in the *communis doctorum opinio* with the tenacity of a popular prejudice. Nevertheless, it is not only vulnerable, but will not stand the test of serious criticism. It is defective in various respects; to the extent that it contains "I" and "my"; to the extent that it contains the conception of "given"; and lastly to the extent that it speaks "only of sensations."

In point of fact, it is only in the case of the first-created Adam, just manufactured out of clay and *for the first time* seeing, again with eyes opened for the *first* time, the landscape of paradise with all its attributes, that such a statement could be made. Any empirical subject *always* goes beyond the bounds of "pure" sensual "raw material"; his experience, representing the result of the influence of the external world on the knowing subject in the process of his practice, stands on the shoulders of the experience of other people. In his "I" there is *always* contained "we." In the pores of his sensations there already sit the products of *transmitted* knowledge (the external expression of this are speech, language and conceptions adequate to words). In his *individual* experience there are included beforehand society, external nature and history—i.e., social history. Consequently, epistemological Robinson Crusoes are just as much out of place as Robinson Crusoes were in the "atomistic" social science of the eighteenth century.

But the thesis criticised is defective not only from the standpoint of "I," "my," "only sensations." It is defective also from the standpoint of "given." Examining the work of A. Wagner, *Marx* wrote: "The doctrinaire professor represents the relations of man and nature from the very outset not as practical relations—i.e., those founded on action, but as *theoretical* . . . but people never begin under any circumstances with 'standing in theoretical relationship with *objects outside the*

---

[2] *George Berkeley*: "Treatise concerning the Principles of Human Knowledge," vol. i. of Works, ed. Frazer, Oxf., 1871.

[3] *Moritz Schlick*: "Die Wende der Philosophie" in "Erkenntnis," vol. i., No. 1. "Ich bin nämlich überzeugt, dass wir sachlich berechtigt sind, den unfruchtbaren Streit der System als beendigt (N.B.) anzuzehen" (p. 5).

[4] *R. Avenarius*: "Kritik der reinen Erfahrung," v. I., Leipzig, 1888, pp. vii. and viii.

world.' Like other animals, they begin by *eating, drinking,* etc. —i.e., they do not 'stand' in any relationship, but *function actively,* with the help of their actions take possession of certain objects of the outside world, and in this way satisfy their requirements. (Consequently they begin with production.)"[5]

Thus the thesis criticised is incorrect also because it expresses a calmly passive, contemplative point of view, and not an active, functioning point of view, that of human practice, which also corresponds to objective reality. Thus, the far-famed "irrefutable" epistemological "axiom" must fall to the ground. For it is in categorical contradiction to objective reality. And it is in just as categorical contradiction to the whole of human practice; (1) it is individualistic and leads directly to solipsism; (2) it is anti-historical; (3) it is quietist. Therefore it must be rejected with all decisiveness.

Lest there should be any misunderstanding: we entirely adopt the standpoint that sensuality, sensual experience, etc., having as their source the material world existing outside our consciousness, constitute the point of departure and beginning of cognition. It was just from this that began the philosophical rebellion of *Feuerbach* against the yoke of the idealistic abstractions and panlogism of *Hegel.* Of course, individual sensations are a fact. But historically there is *no* absolutely unmixed individual sensation, beyond the influence of external nature, beyond the influence of other people, beyond the elements of mediated knowledge, beyond historical development, beyond the individual as the *product* of society—and society in active struggle against nature. And in the "axiom" under consideration, what is important is its logical "purity." If the latter disappears, the whole "axiom" disappears. For this reason the arguments which we put forward are *actual* arguments.

From the above it can already be seen what a *vast role* the problem of theory and practice plays from the standpoint of *epistemology.*

We pass now to the consideration of this theme.

First of all, it should be noted that both theory and practice are the *activity* of social man. If we examine theory not as petrified "systems," and practice not as finished products—i.e., not as "dead" labour petrified in things, but *in action,* we shall have before us two forms of labour activity, the *bifurcation* of labour into intellectual and physical labour, "mental and material," theoretical cognition and practical action. Theory is accumulated and condensed practice. To the extent that it

---

[5] *K. Marx:* "On the book of Adolph Wagner." First published in *Marx and Engels Archives,* vol. v., pp. 387-388, Moscow, 1930. Marx's italics.

generalises the practice of material labour, and is qualitatively a particular and specific continuation of material labour, it is itself qualitatively a special, theoretical practice, to the extent that it is active (cf. e.g., the experiment)—practice fashioned by thought. On the other hand, practical activity utilises theory, and to this extent practice itself is theoretical. In actual fact we have in every class society *divided* labour and, consequently, a contradiction between intellectual and physical labour—i.e., a contradiction between theory and practice. But, like every division of labour, here too it is a living *unity* of opposites. Action passes into cognition. Cognition passes into action. Practice drives forward cognition. Cognition fertilises practice.[6] Both theory and practice are *steps* in the joint process of "*the reproduction of social life.*" It is extremely characteristic that from of old the question has been asked: "How is *cognition* possible?" But the question is not asked: "How is *action* possible?" There is "epistemology." But no learned men have yet thought of inventing some special "praxeology." Yet one passes into the other, and *Bacon* himself quite justifiably spoke of the coincidence of knowledge and power, and of the interdependence of the laws of nature and norms of practice.[7] In this way *practice* breaks into the theory of cognition, theory includes practice, and *real* epistemology, i.e., epistemology which bases itself upon the unity (not the identity!) of theory and practice, includes the practical criterion, which becomes the criterion of the *truthfulness of cognition.*

The relative social disruption of theory and practice is a basis for a break between the theory of cognition and practical action, or for the construction of a super-experimental

---

6 "Theoretical capacity begins with the presently existing, given, external and transforms it into its conception. Practical capacity, on the contrary, begins with internal definition. The latter is called *decision*, intention, task. It then trasforms the internal into the real and external—i.e., gives it present existence. This transition from internal definition to externality is called *activity*." "Activity generally is the union of the internal and the external. The internal definition with which it begins, as a purely internal phenomenon, must be removed in its form and become purely external. . . . On the contrary, activity is also *the removal of the external*, as it is given directly. . . . The form of the external is changed. . . ." (*G. V. F. Hegel:* "Introduction to Philosophy," sections 8 and 9.)

7 *Francis Bacon*: "Philosophical Works," ed. J. M. Robertson, London, 1905. "Human knowledge and human power meet in one; for when the cause is not known the effect cannot be produced. Nature to be commanded must be obeyed; and that which in contemplation is as the cause is in operation as the rule" (p. 259.) *Franc. Baconis de Verulamio*: "Novum Organum Scientiarum." Apud Adrianum Wijngaerum et Franciscum Moiardum, 1645, p. 31. "Scientia et Potentia Humana in idem coincidunt, quia ignoratio causæ destituit effectum. Natura enim non nisi parendo vincitur; et quod in Contemplatione instar causæ est, id in Operatione instar Regulae est."

theory as a skilled free supplement to the usual and earthly forms of human knowledge.[8] *Hegel* has the unity of theory and practice in a particularly idealistic form (unity of the theoretical and practical *idea* as cognition),[9] unity which overcomes the onesidedness (Einseitigkeit) of theory and practice, taken separately, unity "precisely *in the theory of cognition*."[10] In *Marx* we find the *materialistic* (and simultaneously dialectical) teaching of the unity of theory and practice, of the *primacy* of practice and of the *practical criterion of truth* in the theory of cognition. In this way Marx gave a striking philosophical synthesis, in face of which the laboured efforts of modern pragmatism, with its theological and idealistic contortions, its super-artificial and laborious constructions of fictionalism, etc., seem but childish babble.

The interaction between theory and practice, their unity, develops on the basis of the *primacy of practice*. (1) *Historically*: the sciences "grow" out of practice, the "production of ideas" differentiates out of the "production of things"; (2) *sociologically*: "social being determines social consciousness," the practice of material labour is the constant "force motrice" of the whole of social development; (3) *epistemologically*: the practice of influence on the outside world is the primary "given quality." From this follow extremely important consequences. In the exceptionally gifted "theses" of Marx on *Feuerbach*, we read:

> "Die Frage, ob dem menschlischen Denken gegenständliche Wahrheit zukomme - ist keine Frage der Theorie, sonden eine *praktische* Frage. In der Praxis muss der Mensch die Wahrheit, d.h. Wirklichkeit und Macht, Diesseitigkeit seines Denkens beweisen. Der Streit über die Wirklichkeit oder Nichtwirklichkeit des Denkens, das von der Praxis isoliert ist—ist eine rein *scholastiche* Frage." (2nd Thesis.)
>
> "Die Philosophen haben die Welt nur verschieden

[8] Cf. *Marx and Engels*: "Feuerbach (Idealistic and Materialist Standpoint)," Marx and Engels Archives, vol. i., p. 221: "Division of labour becomes a real division of labour only when a division of material and spiritual labour begins. From that moment consciousness *may* in reality imagine that it is something other than the consciousness of existing practice. From the moment that consciousness begins really to imagine something, without imagining something real, from that time onwards it finds itself in a position to emancipate itself from the world and proceed to the formation of 'pure theory,' theology, philosophy, morality, etc."

[9] "Die Idee als Erkennen, welches in der gedoppelten Gestalt der theoretischen und der praktischen Idee erscheint." (*Hegel*: "Wissenschaft der Logik," 391 (vi., sec. 215).

[10] *Lenin:* Abstract of "The Science of Logic," Lenin Review, vol. ix., 6. 270.

interpretiert; es kommst darauf an, sie zu verändern." (11th Thesis.)

The problem of the external world is here put as the problem of its transformation: the problem of the cognition of the external world as an integral part of the problem of transformation: the problem of theory as a practical problem.

Practically—and, consequently, epistemologically—the external world is "given" as the object of active influence on the part of social, historically developing man. The external world has its history. The relations growing up between subject and object are historical. The forms of these relations are historical. Practice itself and theory, the forms of active influence and the forms of cognition, the "modes of production" and the "modes of conception," are historical. The question of the *existence* of the external world is categorically superfluous, since the reply is already evident, since the external world is "given," just as practice itself is "given." Just for this reason in practical life there are no seekers after solipsism, there are no agnostics, no subjective idealists. Consequently epistemology, *including* praxiology, epistemology which *is* praxiology, must have its point of departure in the reality of the external world: not as a fiction, not as an illusion, not as a *hypothesis*, but as a basic fact. And just for this reason Boltzmann[11] "declared with every justification that the premise about the unreality of the external world is "die grösste Narrheit, die je ein Menschengehirn ausgebrütet hat": it is in contradiction to *all the practice of humanity*. Whereas *E. Mach*, in his "Analysis of Sensations," considers that from the scientific (and not the practical) standpoint the question of the reality of the world (whether it exists in reality, or whether it is an illusion, a dream) to be impermissible, since "even the most incongruous dream is a fact no worse than any other."[12] This "theory of cognition" acquired from Vaihinger[13] a demonstrative character, as he erected *fiction* into a principle and "system" of cognition. This peculiar somnambulistic epistemology was foreseen in his day by *Calderon*:[14]

¿Què es la vida? Un frenesí:
¿Que es la vida? Una illusion,
Una sombra, una ficcion,

[11] *Boltzmann:* "Populäre Schriften," 905.

[12] *E. Mach:* "Analyse der Empfindungen."

[13] *R. Vaihinger*: "Die Philosophie des Als Ob. System der theoretischen, praktischen und religiösen Fiktionen der Menscheit auf Grund eines idealistischen Positivismus," Berlin, 1911, p. 91. "Das die Materie eine solche *Fiktion* sei, ist heutzutage eine allgemeine Ueberzeugung der denkenden Köpfe."

[14] *Calderon*: "La Vida es Sueño." Las Comedias del célebre poeta español *Don Pedro Calderon de la Barca*. Zuickavia, Libreria de los hermanos Schumann, 1819.

Y el mayor bien es pequeño,
Que toda la vida es sueño,
Y los sueños sueño son.

Practice is an active break-through into reality, egress beyond the limits of the subject, penetration into the object, the "humanising" of nature, its alteration. Practice is the refutation of agnosticism, the process of transforming "things in themselves" into "things for us," the best proof of the adequacy of thought, and of its *truth*—understood historically, as a process. For, if the objective world is changed through practice and according to practice, which includes theory, this means, that practice verifies the truth of theory; and this means that we *know* to a certain extent (and come to know more and more) objective reality, its qualities, its attributes, its regularities.

Therefore the *fact of technology,* as *Engels* already remarked in "Anti-Dühring,"[15] confutes Kantian agnosticism—that "paltry doctrine," in the words of Hegel.[16] If *K. Pearson* in a "Grammar of Science" modernises the well-known cave of Plato, replacing it by a telephone exchange, and the pale shades of the Platonic ideas by telephone signals, he thereby demonstrates his own conception of the passively-contemplative character of cognition. The real subject—i.e., social and historical man— is not in the least like either *Karl Pearson's* telephonist or the observer of the Platonic shades. He likewise does not in the least resemble that stenographer, inventing "convenient" signs in shorthand, into whom the philosophising mathematicians and physicists desire to transform him (*B. Russell, Wittgenstein, Frank, Schlick,* and others). For he is actively transforming the world. He has changed the face of the whole of the earth. Living and working in the biosphere,[17] social man has radically remoulded the surface of the planet. The physical landscape is ever more becoming the seat of some branch of industry or agriculture, an artificial material medium has filled space, gigantic successes of technique and natural science confront us, the radius of cognition, with the progress of exact apparatus of measurement and new methods of research, has grown extremely wide: we already weigh planets, study their chemical composition, photograph invisible rays, etc. We foretell objective changes in the world, and we *change* the world. But this is unthinkable without real knowledge. Pure symbolism, stenography, a system of signs, of fictions,

---

15 *F. Engels*: Herrn Eugen Dührings Umwälzung der Wissenschaft."

16 "That we do not know realities, and that it has been granted us to know only accidental and passing—i.e., *paltry* phenomena—that is the *paltry* doctrine, which has made and is making the loudest noise, and which now predominates in philosophy." *Hegel*: "Encyclopaedia of Philosophic Sciences," Part I., Speech of Oct. 22, 1818.)

17 See *V. Vernadsky, Member of Academy*: *The Biosphere.* Leningrad, 1926. (Russian.)

cannot serve as an instrument of *objective* changes, carried out by the subject.[18]

Cognition, considered historically, is the more and more adequate *reflection* of objective reality. The fundamental criterion of the correctness of cognition is therefore the criterion of its adequateness, its *degree of correspondence* to objective reality. The *instrumental* criterion of truth is not in contradiction to this criterion, but *coincides* with it, if it is only a question of an instrument for the practice of social man, transforming the objective world (Marx's "revolutionäre Praxis," Engels' "umwälzende Praxis"), and not of the individual "practice" of any philistine in a beershop. Therefore the "instrumental criterion" of pragmatism (*Bergson*, close to pragmaticism; *W. James* and others) must be rejected with all decisiveness. *James* includes as practice, prayer, the "experience" of religious ecstasy, etc.; doubting the existence of the material world, he does not doubt at all the existence of God, like, by the way, many other adherents of so-called "scientific thought" (A. S. Eddington, R. A. Millikan, etc.).[19] The criterion of *economy of thought* can in no way serve as a criterion, since the economy itself can only be established *post factum*: while taken in isolation, as a bare principle of cognition in itself, it means the a priori liquidation of the complexity of thought—i.e., its deliberate incorrectness. In this way "economy" is transformed into its very opposite. "Man's thinking is only 'economic' when it *correctly* reflects objective reality, and the criterion of this correctness is practice, experiment, industry."[20]

---

18 Characteristic of the modern physicists and mathematicians is the following opinion of *Ph. Frank*: "Wir sehen: bei keiner Art von solchen Problemen handelt es sich darum, eine 'Uebereinstimmung zwischen gedanken und Objekt," wie die Schulphilosophie sagt, hervorzubringen, sondern immer nur um die *Erfindung eines Verfahrens*, das geignet ist, mit Hilfe eines geschickt gewählten Zeichensystems Ordnung in unsere Erlebnisse zu bringen und dadurch uns ihre Beherrschung zu erleichtern." (*Ph. Frank*: "Was bedeuten die gegenwärtigen physikalischen Theorien für die allgemeine Erkennt -nislehre?" in "Erkenntnis," vol. i., pp. 2-4; pp. 134-135).

19 "God is real, since he produces real effects" (517). "I believe the pragmatic way of taking religion to be the deeper way. . . . What the more characteristically divine facts are, apart from the actual inflow of energy in the faith-state and the prayer-state, I know not. . . . But the overbelief on which I am ready to make my personal venture is that they exist" (519). *William James*: "The Varieties of Religious Experience," London, 1909. Cf. also "Pragmatism," p. 76. *Study* (loc. cit. 65, footnote) rightly observes: "Er (Vaihinger, N.B.), verurteilt den Pragmatismus *meretrix theologorum*. Ich hatte den Pragmatismus 'die Leib—und Magenphilosophie des banalen Nützlichkeitsmenschen genannt'."

20 *V. I. Lenin*: "Materialism and Empiriocriticism," Works, Eng. ed., vol. xiii.

We see, consequently, that modern capitalist theories of cognition either do not deal with the question of practice altogether (Kantianism: cf. *H. Cohen*: "Logik der reinen Erkennt-nis," 1902, p. 12." Wir fangen mit dem Denken an. Das Denken darf keinen Ursprung haben ausserhalb seiner selbst"), or treat of practice in the Pickwickian sense, tearing it away from the material world or from "the highest" forms of cognition (pragmatism, conventionalism, fictionalism, etc.). The only true position is held by *dialectical materialism,* which rejects all species of idealism and agnosticism, and overcomes the narrowness of mechanical materialism (its ahistorism, its anti-dialectical character, its failure to understand problems of quality, its contemplative "objectivism," etc.).

II.—*Theory and Practice from the Sociological Standpoint. Historical Forms of Society and the Connection of Theory and Practice.*

Dialectical materialism, as a method of cognition applied to social development, has created the theory of *historical materialism.* The usual conception of Marxism is that of a variety of the mechanical, natural-scientific materialism typical of the teachings of the French encyclopædists of the xviii. century or *Büchner-Moleschott.* This is fundamentally wrong. For Marxism is built up entirely on the idea of *historical development,* foreign to the hypertrophied rationalism of the encyclopædists.[21] The question of *theory in general* must be put as follows from what is said above—from the standpoint of *social theory*—*i.e.*, the standpoint of *sociology* and *history.*

At the present time all scientists more or less acquainted with the facts, and all research workers, recognise that *genetically* theory grew up out of practice, and that any branch of

---

21 It is characteristic that, in spite of this, the numerous "refutations" of Marxism systematically begin with the premise of the mechanical character of dialectical materialism and its sociological side (the theory of historical materialism). Cf. *N. N. Alexeyev*: "The Social and Natural Sciences in the Historical Interrelation of their Methods. Part I. The Mechanical Theory of Society. Historical Materialism." Moscow, 1912. Other attempts at a deeper criticism are founded on a poor acquaintance with the subject, though their name is legion.

science has, in the long run, its practical roots.[22] From the standpoint of social development, science or theory is the continuation of practice, but—to adapt the well-known remark of *Clausewitz*—"by other means." The function of science, in the sum total of the process of reproduction of social life, is the function of orientation in the external world and in society, the function of extending and deepening practice, increasing its effectiveness, the function of a *peculiar* struggle with nature, with the elemental progress of social development, with the classes hostile to the given socio-historical order. The idea of the self-sufficient character of science ("science for science's sake") is naïve : it confuses the *subjective passions* of the professional scientist, working in a system of profound division of labour, in conditions of a disjointed society, in which individual social functions are crystallised in a diversity of types, psychologies, passions (as Schiller says: "Science is a goddess, not a milch cow"), with the objective *social rôle* of this kind of activity, as an activity of vast *practical* importance. The fetishising of science, as of other phenomena of social life, and the deification of the corresponding categories is a perverted ideological reflex of a society in which the division of labour has destroyed the visible connection between social function, separating them out in the consciousness of their agents as absolute and sovereign values. Yet any—even the most abstract—branch of science has a quite definite vital importance in the course of historical development. Naturally, it is not a question of the *direct* practical importance of *any* individual principle—e.g., in the sphere of the theory of numbers, or the doctrine of quantities, or the theory of conditioned reflexes. It is a question of systems as a whole, of appropriate activity, of *chains* of scientific truths, representing in the *long run* the theoretical ex-

[22] Cf. on mathematics among the Babylonians, Egyptians, Greeks, Romans, Chinese, Indians, etc. *M. Kantor*: "Vorlesungen über die Geschichte der Mathematik." Leipzig. Trubner, 1903, vol. i., 3rd ed. Cf. also *F. J. Moore*: "History of Chemistry." *Otto Wiener*: "Physics and the Development of Culture." *R. Eisler*: "Geschichte der Wissenschaften." *A. Bordeaux*: "Histoire des sciences physiques, chimiques, et géologiques au xix. siècle," Paris et Liège, 1920. "It is necessary to study the *successive development* of individual branches of natural science. First *astronomy*—already from year to year absolutely necessary for pastoral and agricultural peoples. Astronomy can develop only with the help of *mathematics*. Consequently, it became necessary to study the latter, too. Further, at a certain stage of development and in certain countries (the raising of the water level for irrigation purposes in Egypt), and particularly together with the origin of towns, large buildings, and the development of handicrafts, there developed *mechanics* also. Soon it also became necessary for *shipping* and the *art of war*. . . . Thus from the very beginning the origin and development of Sciences are conditioned by production." (*F. Engels*: "Dialectics of Nature. Dialectics and Natural Science." Marx and Engels Archives, II., p. 69.)

pression of the "struggle with nature" and the social struggle. Active relationship with the external world, which at the purely animal stage of human development presupposes the natural organs of man, as a variety of *hominis sapientis*, is replaced by relationship through the medium and with the help of the "continuation of those organs," i.e., with the help of the "productive organs of social man" (Marx), the implements of labour, and systems of social technique. At first this system is really the "continuation" of the organs of the human body.[23] Later it becomes complicated, and acquires its own principles of movement (e.g., the circular motions of modern machinery). But at the same time there develops historically also *orientation* in the external world, again with the help of *artificial instruments* of cognition, instruments of "spiritual" labour, extending a gigantic number of times the sphere of action of the natural organs of the body and the instruments of orientation. Micro-balances, the water-level, seismographs, the telephone, the telescope, the microscope, the ultra-microscope, the chronoscope, the Michelson grating, electrical thermometers, bolometers, the photo-electrical element of Elster and Geitel, galvanoscopes and galvanometers, electrometers, the apparatus of Ehrenhaft and Millikan, etc., etc.—all these immeasurably widen our natural sensual capacities, open new worlds, render possible the victorious advance of *technique*. It is a piece of historic irony, at the expense of the greatly multiplied agnostics who completely fail to understand the value of transmitted knowledge,[24] and reduce the whole process of cognition to the production of *tautology*, that precisely the *electrical* nature of matter is the "last word" of science: since it is just the "electrical feeling" which we lack. "Yet the whole world of electricity was discovered to us none the less by means of the application of artificial organs of sensation."[25] Thus there have proved to be *historically variable* both the "organs of sensation" and the so-called "picture of the world," *verified by the gigantic practice of modern humanity as a whole*, a "picture of

23 Cf. *Marx*: "Capital," Eng. ed., vol. i., p. 158: "Thus Nature becomes one of the organs of his activity, one that he annexes to his own bodily organs, adding stature to himself in spite of the Bible." Cf. also *Ernst Kapp*: "Grundlinien einer Philosophie der Technik." Braunschweig, 1877, pp. 42 et seq.

24 ". . . Vielmehr glauben wir, dass nur die Beobachtung uns Kenntnis vermittelt von den Tatsachen, die die Welt bilden, während *alles Denken nichts ist als tautologisches Umformen*." (*Hans Hahn*: "Die Bedeutung der wissenschaftlichen Weltauffassung, insbesondere für Mathematik und Physik" in "Erkenntnis," I., Nos. 2-4, p. 97, 1930. The group of empiriocritics fail to understand that the product of perceptive activity is *qualitatively* different from sensual "raw material," just as the completed locomotive is qualitatively different from its metallic parts, even though 'made' out of them.

25 *O. Wiener*, Op. cit., p. 41.

the world" much more adequate to reality than all its predecessors, and *therefore* so fruitful for practice.

And so man is historically given as *social* man (in contradistinction to the enlightened Robinsons of *Rousseau*, "founding" society and history like a chess club, and with the help of a "contract." This social man, i.e., human society, in order to live, must *produce*. *Am Anfang war die Tat* (in contrast to the Christian Logos: "In the beginning was the Word"). Production is the real starting point of social development.[26] In the process of production there takes place a "metabolism" (Marx) between society and nature. In this process, *active* on the part of historical and social man, a *material* process, people are in definite relationship one with another and with the means of labour. These relations are historical, their totality constitutes the *economic structure of society*. It is also a historic variable (in contradistinction to the theories of "society generally," "eternal society," "ideal society," etc.). The economic structure of society (the "mode of production") includes, above all, the relationship between *classes*. On this basis there grows up the "superstructure": political organisations and State power, moral norms, scientific theories, art, religion, philosophy, etc. The "mode of production" determines also the "mode of conception": theoretical activity is a "step" in the reproduction of social life; its material is furnished by experience, the breadth of which depends on the degree of power over the forces of nature, which is determined, in the long run, by the development of productive forces, the productivity of social labour, the level of technical development. Stimuli proceed from the tasks set by practice; the forming principles, the "mode of conception" in the literal sense, reflect the "mode of production," the socio-class structure of society and its complex requirements (the idea of rank, authority, the hierarchy and the personal God in feudal society; the idea of the impersonal force of fate, of the elemental process, of the impersonal God in capitalist commodity-society," etc.). The prevailing conceptions are those of the ruling class, which

---

[26] This is not a secret for some modern physicists either. "The physical conditions of existence are more fundamental than the aesthetic, moral, or intellectual. A child must be fed before it can be taught. A certain standard of living above that of animals is a preliminary condition for the development of any of the special qualities of human beings." (*Frederic Soddy*: "Science and Life," London, J. Murray, 1920, p. 3.)

is the *bearer* of the given mode of production.[27]

But, just as development in natural history changes the forms of biological species, the historical development of society, with the movement of productive forces at its foundation, changes the socio-historic forms of labour, "social structures," "modes of production," together with which there changes the whole ideological superstructure, up to and including the "highest" forms of theoretical cognition and reflective illusions. The movement of productive forces, the contradiction between them and the historic forms of social labour are, consequently, the cause of the *change* in these forms, realised through class struggle (to the extent that we are speaking of class societies) and the blowing up of the out-of-date social structure, transformed from "a form of development" to "fetters on development." In this way the practice of *material labour* is the basic motive force of the entire process as a whole, the practice of the *class struggle* is the critical-revolutionary practice of social transformation ("criticism by

---

[27] The fashionable German philosopher and author of "Christian-prophetic" "Socialism," *Max Scheler*, while carrying on a desperate struggle against Marxism, borrows from the latter a number of basic principles, producing as a consequence a perfectly intolerable cacophony of *motifs*. To illustrate the influence of Marxism on this Catholic philosopher, we quote the following passage from his large work, "Die Wissensformen und die Gesellschaft" (Leipzig, mcmxxvi), pp. 204-205: "So ist es nicht unrichtig, dass selbst sehr formale Arten des Denkens und der Wertnehmung klassenmässig verschieden geartet sind—freilich nur in Gesetzen der grossen Zahl der Fälle, da ja jeder die Bindung seiner Klassenlage prinzipiell überwinden kann. Zu solchen klassenmässig bestimmten formalen Denkarten rechne ich beispielweige folgende:—

1. Wert-prospektivismus des Zeitbewussteins—Unterklasse; Wert-retrospektivismus-Oberklasse.
2. Werdensbetrachtung-Unterklasse: Seinsbetrachtung-Oberklasse.
3. Mechanische Weltbetrachtung-Unterklasse; teleologische Weltbetrachtung-Oberklasse.
4. Realismus (Welt vorwiegend als 'Widerstand')-Unterklasse; Idealismus-Oberklasse (Welt vorwiegend als 'Ideenreich').
5. Materialismus-Unterklasse; Spiritualismus-Oberklasse.
6. Induktion, Empirismus-Unterklasse; Aprioriwissen, Rationalismus-Oberklasse.
7. Pragmatismus-Unterklasse; Intellektualismus-Oberklasse.
8. Optimistische Zukunftsansicht und pessimistische Retrospektion-Unterklasse. Pessimistische Zukunftsaussicht und optimistische Retrospektion, 'die gute alte Zeit'-Oberklasse.
9. Widersprüche suchende Denkart oder 'dialektische' Denkart-Unterklasse; identitätssuchende Denkart-Oberklasse.
10. Milieu-theoretisches Denken-Unterklasse; nativistisches Denken-Oberklasse."

This original table is extremely schematic and unhistorical, but it contains individual elements of truth. However, this truth does not prevent *Scheler* from standing pat on the side of the "Oberklasse" and launching into the wilds of appropriate religious metaphysics.

weapons" which takes the place of the "weapon of criticism"), the practice of *scientific cognition* is the practice of material labour continued in particular forms (natural science), of administration and the class struggle (the social sciences). The "class subjectivism" of the *forms of cognition* in no way excludes the objective "significance" of cognition: in a certain measure cognition of the external world and social laws is possessed by every class, but the specific methods of conception, in their historical progress, *variously* condition the process of the *development* of the adequateness of cognition, and the advance of history may lead to such a "method of conception" as will become a fetter upon cognition itself. This occurs on the eve of the destruction of the given mode of production and its class promoters.

It is from this *historico-materialist* angle that we should also approach the exceptionally complicated question of *the interrelations between the theoretical* ("*pure*") *and applied sciences*. Here there is a considerable number of various solutions: (*a*) to take as a criterion the difference between causal theoretical series ("Naturgesetz," law) and teleological, normative series (rule, system of rules, prescriptions);[28] (*b*) to take as a criterion distinction according to objects—the "pure" sciences study the natural surroundings given to man: the applied sciences the artificial surroundings (machines, transport technique, apparatus, raw materials, etc.);[29] (*c*) to take as criterion time (the "pure" sciences work with a long period in view, forestalling developments, the applied serve "the needs of the moment");[30] (*d*) to take as criterion, lastly, the degree of generality ("abstractness") of the particular science.

On this subject it is necessary to remark (*a*) on the first criterion: "sciences" teleologically set forth at bottom are not sciences, but arts (Künste). However, any system of norms (we have not here in mind ethics and the like) depends upon a system of objective laws, which are either covertly understood or directly set forth as such. On the other hand, the sciences in the particular sense of the word ("pure sciences") are not "pure," since the *selection of an object* is determined by aims which are practical in the long run—and this, in its turn, can and must be considered from the standpoint of the causal *regularity* of social development.[31]

---

28 Cf. *E. Husserl*: "Logical Researches." Cf. *M. Lomonosov*: "On the Value of Chemistry," Works (St. Petersburg, 1840), iii., p. 1.

29 Cf. *Paul Niggli*: Reine und angewandte Naturwissenschaft. "Die Naturwissenschaften." 19. Jahrgang, Heft I.

30 Cf. *W. Ostwald*: "Der energetische Imperátiv," I. Reihe, Leipzig 1912, pp. 46, 53.

31 The attempts, recently still fashionable, of the school of *H. Rickert* to dig an impassable abyss between the social and natural sciences logically rely upon the naïve conception that in the natural sciences, as opposed to the social, there is no "relation to values." This

(*b*) On the second criterion: engineering, for example, may be set forth as a "pure" study—i.e., theoretically, without norms, without constructive rules; however, usually in its enunciation we also have a teleological and normative element. The same has to be said, e.g., of the resistance of materials, the science of staple commodities, and so forth. This is not an accident, for here the *object* itself (" the artificial surroundings") is material *practice*.

(*c*) On the third criterion: a vividly practical task may also be "protracted" (e.g., the problem of aeronautics, as it stood for a number of *centuries*, or—at the present time—the transmission of energy from a distance), a task which *always* has its "purely theoretical" equivalent as well.

(*d*) On the fourth criterion: a very concrete science may also be "purely theoretical," since knowledge has broken up into a number of rivulets, and has become extremely specialised. It would hardly come into anyone's head, for example, to classify the Japhetic theory of language among the applied sciences, although it also, of course, is bound up with a number of the most important practical tasks. (Here we should also note the relativity of the conceptions of concrete and abstract.)

And so, apparently, all the definitions are defective. The most accurate definition is the division according to the characteristic of causal and teleological series. However, here too we see obvious defects from the standpoint of actual relation-

---

"relation to values" exists in the natural sciences as well, so far as *selection of an object* is concerned. However, *teleology* must be *driven out* of science, as a system of theoretical principles discovering objective regularities, and this applies *equally* to the social and the natural sciences. The raison-d'être of the Rikkertian view for the bourgeoisie, however, is that its social science is rapidly declining to scientific non-existence, changing more and more into the simple *apology* of the capitalist system, which for the *Rickerts* undoubtedly has a most outstanding "value." As for the other distinction of "principle" made by Rickert (the historical character of the social sciences and the non-historical character of the natural), it relies upon an extreme narrowness of outlook, which takes note of the historical evolution of some social phenomena, but does not see the *history of nature*. At the present time a new school is arising in place of Rickert-Dilthey—M. Weber, O. Spann, W. Sombart—which proclaims the impossibility of the perception of external nature ("the essence of things") and the full possibility of the perception of the "sense" of social phenomena, *Sombart* moreover maintaining that the natural sciences have practical value, while the social sciences cannot have any practical application. Truly modern bourgeois science is beginning to walk on its head! Cf. *Sombart*: Die drei Nationalökonomien v. Geschichte und System der Lehre von der Wirtschaft. Duncker und Humblot, 1930.

ship. But all these defects of logical definition reveal the objective dialectics of reality: contradictions arise here because there is an objective contradiction between theory and practice, and at the same time their unity; there is their difference, as opposite poles of human activity, and at the same time their interpretation; there is their separate existence as functions, as branches of divided social labour, and at the same time their unitary existence, as steps in the joint "production of social life." Under the cover of the difficulty of the exact demarcation of the applied and theoretical sciences beats the dialectics of the relationship between theory and practice, the *passing* of one into the other: which does not fit—and cannot fit—into the framework of school-logic and academical-pedantic definitions. In reality we have a whole chain of various theoretical sciences, linked up by internal connections ("the *classification of sciences,* of which each analyses a separate form of motion or a number of interconnected forms of motion which pass into one another, is also a classification or hierarchy of these very forms of motion according to the order inherent in them: and just in this lies its significance.")[32] These sciences are born out of practice, which first sets itself "technical" tasks: the latter require, in their turn, the solution of "theoretical" problems, problems of the first, second, etc., order, a special (relative) logic of motion being thereby created. Practice in this way grows into theory: the sought-for rule of action is transformed into the search for the law of objective relationship: there arise innumerable knots and interlacings of problems with their solutions: these, in their turn, sometimes fertilise a number of hierarchically lower branches of science, and through technology penetrate into technique—consequently, into the direct practice of material labour, transforming the world. Here law becomes transformed into a rule of action, the percipient decision is verified by that action, orientation in the surroundings becomes the alteration of those surroundings, the intellect is immersed in the will, theory once again reverts to the form of practice. But this metamorphosis has as its final result by no means a simple repetition of the previous cycle of practice, since practice becomes practice on a more powerful and qualitatively altered basis.

The problem of the "pure" and "applied" sciences, reflecting and expressing the problem of theory and practice, is not however a purely logical problem. It is itself a problem of history, and a problem of transforming *historical practice.* The acuteness of the problem in the innermost recesses of the

[32] *F. Engels*: "Dialectics of Nature," pp. 31-33. See also *Hegel*: "Phenomenology of the Spirit" (St. Petersburg, 1913, p. 112): "Symptoms must not only bear an essential relationship to cognition, but must also be essential definitions of things, so that the artificial system must be in conformity with the system of nature itself, and express only that system."

capitalist order, and even the posing of the problem itself, is the theoretical expression of the real separation, fixed in terms of profession and class, and rupture between theory and practice—a rupture, naturally, relative and not absolute. This rupture, consequently, is a historical phenomenon: it is bound up with a definite historico-economic formation, with a definite historically transitory "mode of production," with the bifurcation of labour into intellectual and physical labour, with the polarisation of classes. It may therefore be said with every justification that socio-economic formations ("modes of production," "economic structures") differ from one another also *in the particular character of the relationship between theory and practice*. And in fact, in the theocratic state of Ancient Egypt there were elements of a natural centralised planned economy; knowledge (theory) was most closely connected with practice, since it was expediently directed towards practice. But this connection was of a special type. Knowledge was inaccessible to the mass of workers: their practice for *them* was blind, and knowledge was surrounded with an aureole of dread mystery. In this sense there was a vast *rupture* between theory and practice. If we take for comparison the epoch of industrial capitalism, the epoch of the flourishing of "economic man," of boundless individualism, of "laissez faire," we see a different picture. On a social scale no one puts forward in an organised fashion either problems of cognition or problems of application of acquired knowledge. The division of labour creates a group of scientists and ideologues, bound up with the ruling class, which in its turn is broken to pieces by competition. The connection between theory and practice is to a considerable extent built up "privately." But the bifurcation of intellectual and physical labour does not disappear: it receives a different expression—a certain degree of "democratisation of knowledge," necessary from the standpoint of technique: the formation of a large stratum of technical and other intelligentsia: the specialisation of science: the creation of high theoretical generalisations, completely remote from the *consciousness* of the mass of practical workers (wage-workers). This is another type of connection.[33] Its inevitable consequence is the

[33] A number of other examples might be quoted. *Moore*, in his "History of Chemistry," already quoted, writes of the Greek philosophers: "They lacked direct acquaintance with chemical transmutations. Owing to their social position they were deprived of direct contact with those who might have communicated practical information to them, while the general spirit of the age forced them to despise experiment, equally with physical work. Only pure thought was considered worthy of a philosopher" (p. 2). "The slow progress of science in antiquity is explained by the dissociation of theory from practice. There existed no contact between those who worked and those who thought" (pp. 9, 10). Cf. also *Hermann Diels*: "Wissenschaft und Technik bei den Hellenen" in "Antike Technik." (Trubner, Leipzig & Berlin, 1920), pp. 21 et seq. Cf. with this observation *Marx* on Aristotle in "Capital," vol. I.

abstract and impersonal fetishism of science (science for science's sake), the disappearance of the social self-consciousness of science, etc. Modern capitalism reproduces this anarchy on the new and more powerful basis of trustified industrial complexes and the corresponding scientific organisations. But it cannot either discover a scientific synthesis, or attain the self-knowledge of science, or achieve its organisation, or its *fusion with practice*. These problems, which are poignantly felt, lead already *beyond the boundaries of capitalism.*

III.—*Theory and Practice of the U.S.S.R. and the Empirical Test of Historical Materialism.*

It follows from all the foregoing that the question of theory and practice is simultaneously both a theoretical and a practical question: that both theory and practice, and likewise the forms of combination of theory and practice, are bound up with a definite historical order of society, its development, its "motion." Therefore it is beyond all doubt that a particularly stormy course of social life (a revolution) and a new social order (Socialism im Werden) are of exceptional interest from the standpoint of the problem we are considering.

All knowledge is tested in practice, by experience. The same has to be said of the systematised knowledge, of theory, theoretical tendency, "doctrine." It is relevant here to record, first of all, that *Marxism*, weighed in the balance of history, has been verified therein in the most varied directions. Marxism foretold the war; Marxism foretold the period of revolutions and the whole character of the epoch we are going through; Marxism foretold the dictatorship of the proletariat and the rise of a Socialist order; even earlier had been brilliantly justified the theory of the concentration and centralisation of capital, etc. The Revolution has proved *the great destroyer of fetishes*, laying bare the fundamental links and interdependences of society in their real significance. The *State* appeared to bourgeois science now as a distinct organism (even up to the point of determining its sex), now as a fantasy, now as an expression of the "Absolute Spirit," now as the universal organisation of the popular will, etc. The Revolution has destroyed one State and built up another: it has *practically* invaded this sphere of reality, and has ascertained the component parts of the State, and its functions, and its personnel, and its "material appendages," and its class significance, and its significance from the standpoint of economics. The Revolution has *completely* confirmed the theoretical teaching of *Marx* on the State. The same has happened to the norms of law, with "law" itself: juridical fetishism has burst into atoms. *Morality*, which found its "theoretical justification" in the categorical imperative of *Kant,* and which reached its highest stage of deification, disclosed itself to be a system of relative historical norms, with a quite earthly, quite social, and quite historical origin. *Religion,* which is revered as the highest product of human thought, proved to be a cast taken from a society of lords and slaves, a construction on the model of a dualist society, on the model of a hierarchical ladder of domination and exploitation. For this very reason it began rapidly to die out.

But the revolution in *reflective categories,* which was the inevitable result of the *material* revolution, has not yet concluded. We are patently viewing its first phase. Here it is necessary to dwell on some problems in this connection, related to the question of theory and practice.

The capitalist economic order is a system of unorganised elementally developing, and *as a whole* irrational economic life ("anarchy of production," competition, crises, etc.). The Socialist economic order is a system of organised, planned, and anti-exploiter economy, in which little by little there disappears the division between town and country, intellectual and physical labour. Hence follow vast consequences. First of all, it is necessary to note the changes in *the character of social regularity.* The regularity of capitalism is an *elemental regularity,* coming into existence irrespective of (and sometimes against) the will of man (typical examples are the regularity of the industrial cycle, of *crisis,* etc.). This regularity shows itself in the shape of a compulsory law, "like the law of gravity when a house falls on your head."[34].

In relation to the actions of individual persons this regularity is irrational, even though every one of them should act according to all the rules of rational calculation. This irrational current of life is the consequence of the anarchic character of the capitalist structure. The regularity in organised Socialist society is of a different type. It loses (if we are speaking of a process, it *begins to lose*) its elemental character: the future lies ahead as a plan, an aim: *causal* connection is realised through social *teleology*: regularity shows itself not *post factum,* not unforeseen, incomprehensible, blind: it shows itself as "recognised necessity" ("freedom is recognised necessity"), realised through *action organised on a social scale.* Consequently, here is present a different type of regularity, a different relationship between the individual and society, a different relationship between causal and teleological series. In capitalist society the theoretical foreknowledge of the general course of events does not provide the instrument for taking direct control of that course (and there is no subject to set himself such

[34] *K. Marx*: "Capital," vol. i. Cf. also *Engels*: "Ludwig Feuerbach," &c.

a task: society itself is subjectless, blind, unorganised). In Socialist society the theoretical foreknowledge of the necessity can *at once* become a norm of action on the scale of the *whole of society*—i.e., on the scale of "the whole." Thereby is afforded the *possibility* of the fusion of theory and practice, their gigantic social synthesis, historically more and more realised in the measure of the elimination of the rupture between *intellectual and physical labour.*

In the *economic* life of capitalism the elementary social necessity of definite proportions between the branches of production is achieved by means of an elemental fluctuation of prices, in which the law of value expresses itself as the elemental regulator of socio-productive life. In the economic life of Socialism the distribution of resources (means of production and labour power) takes place as a constructive task of a plan. But the plan does not fall from the sky: it is itself the expression of "recognised necessity." Consequently, here (*a*) the tasks of *cognition* expand to a colossal degree; (*b*) this cognition must embrace a huge quantity of problems, and express itself in the work of all branches of science; (*c*) this cognition must become synthetic, for a plan is a *synthesis,* and a scientifically elaborated plan can rely *only* on a synthesis; (*d*) this cognition is directly bound up with practice: it relies on practice, it serves it, it passes into it, for the plan is *active*: it is at one and the same time a product of scientific thought, laying bare causal regularities, and a system of purposes, an instrument of action, the direct regulator of practice and its component part. But the plan of Socialist construction is not only a plan of *economy*: the process of the *rationalisation of life,* beginning with the suppression of irrationality in the economic sphere, wins away from it one position after another: the principle of planning invades the sphere of "mental production," the sphere of science, the sphere of *theory.* Thus there arises here a new and much more complex problem: the problem of the rationalisation not only of the material-economic basis of society, but also of the relations between the sphere of material labour and "spiritual labour," and of relations within the latter—the most striking expression of this is the question of the *planning of science.*[35]

In the *ideological* life of capitalism a certain social necessity of definite proportions (much less definite than in economic life!) between the various branches of ideological labour is regulated to an extremely small extent by the State (the only sphere which is completely regulated is the production and diffusion of religious ideas through the organisation of the

---

[35] For this see: "Proceedings of the 1st Conference on Planning of Scientific Research Work," Moscow, 1931.

State Church.) The regularities of development are here also elemental. Those basic principles which the theory of historical materialism puts forward cannot serve as a standard of action for the ruling class, on the social scale of that action, for the same reason that a capitalist "plan" is unrealisable: a plan is in contradiction to the very structure of capitalism, the prime dominants of its structure and its development. Here, too, the building of Socialism puts the whole problem in a new way. The elemental regularity of interdependences between economy and ideology, between collective economic practice and the multifarious branches of theoretical labour, yield place to a considerable degree to the principle of planning. At the same time, all the basic proportions of the theory of historical materialism are confirmed: one can feel with one's hands, as it were, how the requirements of the rapid and intensive growth of the U.S.S.R. imperiously dictate the solution of a number of technical problems, how the solution of these problems, in its turn, dictates the posing of the greatest theoretical problems, including the general problems of physics and chemistry. One can feel with one's hands how the development of Socialist agriculture pushes forward the development of genetics, biology generally, and so on. It can be observed how the exceptionally insistent need for the study of the natural wealth of the Union broadens the field of geological research, pushes forward geology, geochemistry, etc. And all the poverty of the idea that the "utility" of science means its degradation, the narrowing of its scope, etc., becomes crystal clear and apparent. Great practice requires great theory. The building of science in the U.S.S.R. is proceeding as the conscious construction of the scientific "superstructures": the plan of scientific works is determined in the first instance by the technical and economic plan, the perspectives of technical and economic development. But this means that thereby we are arriving *not only at a synthesis of science, but at a social synthesis of science and practice*. The relative disconnection between theory and practice characteristic of capitalism is being eliminated. The fetishism of science is being abolished. Science is reaching the summit of its social self-cognition.

But the *Socialist* unification of theory and practice is their most *radical* unification. For, gradually destroying the division between intellectual and physical labour, extending the so-called "higher education" to the *whole* mass of workers, Socialism fuses theory and practice in *the heads of millions*. Therefore the synthesis of theory and practice signifies here a quite exceptional increase in the effectiveness of scientific work and of the effectiveness of Socialist economy as a whole. The unification of theory and practice, of science and labour, is *the entry of the masses* into the arena of cultural creative work, and the transformation of the proletariat from an object of culture into its subject, organiser and creator. This revolution in the very foundations of cultural existence is accompanied necessarily by a revolution in the *methods* of science : synthesis presupposes *the unity of scientific method*: and this method is *dialectical materialism,* objectively representing the highest achievement of human thought. Correspondingly is being also built up the *organisation* of scientific work : together with concentrated planned economy there is growing the *organisation of scientific institutions,* which is being transformed into a vast association of workers.[36]

In this way is arising a new society, growing rapidly, rapidly overtaking its capitalist antagonists, more and more unfolding the hidden possibilities of its internal structure. From the standpoint of *world history* the whole of humanity, the whole *orbis terrarum,* has fallen apart into two worlds, two economic and cultural-historic systems. A great world-historic antithesis has arisen : there is taking place before our very eyes the polarisation of *economic systems,* the polarisation of classes, the polarisation of the methods of combining theory and practice, the polarisation of the "modes of conception," the polarisation of *cultures.* The crisis of bourgeois consciousness goes deep, and traces out marked furrows : on the whole front of science and philosophy we have gigantic dislocations which have been excellently formulated (from the standpoint of their basic orientation) by *O. Spann* : the main thing is a war of destruction against *materialism.* This is the great task of culture,[37] in the opinion of the warlike professor, who protests against knowledge without God and knowledge without virtue (Wissen

[36] *Otto Neurath*: "Wege der wissenschaftlichenAuffassung" ("Erkenntnis" vol. i., No. 2-4, p. 124): "In grösstem Stil planmässig gedankliche Gemeinschaftsarbeit ist als Allgemeinerscheinung wohl nur möglich in einer planmässig durchorganisierten Gesellschaft, die mit Hilfe irdisch begründeter Mittel, straff und bewusst die Lebensordnung in Hinblick auf irdisches Glück gestaltet, Soziale Wandlungen sind Präger geistiger Wandlungen." The same author pays a tribute to the materialist conception of history (p. 121), recognising the fact of the true prognoses drawn up by the Marxists. Quite otherwise has been the philosophic evolution of *W. Sombart,* who in his last book writes that Marxism owes its "monstrous" power "ausschlûsslich den in Mystik auslaufenden geschichts-philosophischen Konstruktionen dieser Heilslehre" (*Werner Sombart*: "Die drei Nationalökonomien," p. 32). This charge of mysticism levelled against Marxism is just as stupid as the previously mentioned "essence" and "sense" of the latest "sociology of sense." And bourgeois science is patently beginning to wander in its accusations against the theory of the revolutionary proletariat!

[37] *Dr. Othmar Spann*: "Die Krisis in der Volkswirtschaftlehre," p. 10— :.:. . . so finden wir . . ., dass ein . . . auf Vernichtung hinzielender Kampf gegen . . .sagen wir zuletzt Materialismus jeden Schlages, geführt wurde. Seit der Aufklärung gibt es keine lebenswichtigere Angelegenheit der Kultur."

ohne Gott und Wissen ohne Tugend). In *economic* ideology, under the influence of the crisis of the capitalist system, there has begun the direct preaching of a return "to the pick and the hoe," to pre-machine methods of production. In the sphere of "spiritual culture" the return to religion, the substitution of intuition, "inward feeling," "contemplation of the whole," for rational cognition. The turn from individualist forms of consciousness is patent. It is universal—the idea of "the whole," "wholeness" ("das Ganze," "Ganzheit") in philosophy; in biology (*Driesch* and the Vitalists), in physics, in psychology (Gestaltpsychologie), in economic geography (territorial complexes), in zoology and botany (the doctrine of heterogeneous "societies" of plants and animals), in political economy (the collapse of the school of "marginal utility," "social" theories, the "universalism" of *Spann*), and so on, and so forth. But this turn to the "whole" takes place on the basis of the *absolute* breaking-away of the whole from its parts, on the basis of *idealistic* understanding of the "whole," on the basis of a sharp turn *to religion*, on the basis of the methods of supersensual "cognition." It is not surprising, therefore, that from any scientific hypothesis quasi-philosophic (essentially religious) conclusions are being drawn, and on the extreme and most consistent wing there is openly being advanced the watchword of *a new medievalism*.[38]

In complete opposition to this comprehensible development, young Socialism is arising—its economic principle the maximum of technical economic power, planfulness, development of all human capacities and requirements: its cultural-historical approach determined by the Marxist outlook: against religious metaphysics advancing dialectical materialism: against enfeebled intuitive contemplation, cognitive and practical activism: against flight into non-existent metempirical heavens, the sociological self-cognition of all ideologies: against the ideology of pessimism, despair, "fate," *fatum*, the revolutionary optimism which overturns the whole world: against the complete disruption of theory and practice, their greatest synthesis: against the crystallisation of an "élite," the uniting of the millions. It is not only a new economic system which has been born. A new culture has been born. A new science has been born. A new style of life has been born. This is the greatest antithesis in human history, which both theoretically and practically will be overcome by the forces of the proletariat—the last class aspiring to power, in order in the long run to put an end to all power whatsoever.

---

[38] Cf. *E. Morselli*: Πράττειν, ποιεῖν, Θεωρεῖν in "Rivista di filosofia," vol. xxi., No. 2, "è un ritorno a un nuovo Medio evo che in forme varie agita oggi il pensiero della 'élite' europea" (p. 134). Cf. also *Berdiaeff*: "Un nouveau Moyen Age." Paris, 1927.

PHYSICS AND TECHNOLOGY.

By A. F. JOFFE.

PHYSICS AND TECHNOLOGY.

By A. JOFFE, Sc. D., LL.D.

There is a very close relationship between physics and industry. The truth is that all forms of industry are nothing but various sections of physics or chemistry applied and exploited on a large scale. But it is also true that most conceptions of physics are discovered as the result of consideration of technical problems. The realm of technique is grateful enough to remember the origin of the methods employed by engineers, but the pure scientist usually forgets the manner in which any particular problem found its way into the primers of physics. He begins the history of any problem at the stage where it is already formulated as a scientific problem.

Everybody knows that dynamos and motors owe their existence to Faraday's fundamental discovery of induction, that Maxwell's ideas and Hertz' experiments with electromagnetic waves led to wireless. It is also well known that Lord Kelvin's and R. Clausius' work on thermodynamics laid the basis for the development of thermal technique. The technical bases of the energy and entropy-law clearly formulated by Carnot are often referred to, but the development of thermodynamics since Kelvin is represented as though the thermal technique, metallurgy, and especially the working of steel and alloys, had no influence upon and were independent of the scientific conception of thermodynamic potentials, the theory of phases, and of the *surface state*.

It is instructive to see how the scientific investigation of spark discharge was stimulated by the spark generators in wireless technique, how the wireless valves reacted on the development of our ideas on *electrical emission*, on surface structure, on the theory of atoms, their excitation and ionisation, and led finally to a new theory of metallic states. The growing importance of vacuum techniques and the various applications of photoelements opened up a wide field of investigation which appears to be in a fair way to becoming highly important to our ideas on molecular forces and the mechanism of the transmission of electric charges.

We also note that the most fundamental of problems pass into oblivion when they cease to have technical importance. Electrification by friction was dropped when galvanic cells were invented.

No new types of cell were invented once industry had replaced them by dynamos, despite the fact that the principles of both the friction and the galvanic cells were not fully understood.

Such mutual stimulation is undoubtedly of great benefit both to science and to industry. The unfortunate fact is that it is neither admitted nor even generally desired by scientists. The number of facts we investigate and speculate upon is in reality very limited. Ever since physics made a choice of problems worth studying and continues to bring them within the orbit of a general theory. This limitation has had some unfortunate results. We do not choose the correct theory applicable to a large field of heterogeneous phenomena, but choose the phenomena from the aspect provided by our current theory. We could perhaps avoid many difficulties and disappointments in the theory of light and matter, in statistical mechanics, in the conception of ether, if we adopted both methods for the progress of science.

The physical phenomena presented by the large industries and agriculture are especially adapted to an enlargement of the field of scientific investigations. The great benefit resulting from the smallest improvement, even by a new method of presentation, on the one hand, and the precisely defined conditions and the large scale of the resulting processes on the other hand, are highly favourable to scientific study. Millions of workers, who are familiar with these processes, could be employed in such investigations, and these could be connected up with education and controlled by the scientists of colleges and scientific institutions. We come up against a problem which seems to promise to open up new roads to the progress of science, but those roads cannot be pursued except in a land of Socialism, such as we are trying to build in the Union of Soviet Socialist Republics.

If the relation between science and industry were clearly understood we could expect that science would consciously prepare a basis for the development of technique. There is however no sign of investigations being directed to a solution of the fundamental difficulties of technique. I shall specify a few of the problems forgotten by physicists yet of importance to technique.

1. A reversible oxydation of coal could three or four times increase the energy available for technical purposes.

2. The primary source of all energy, the sun, is exploited only to a ridiculous extent, to a small number of waterfalls. We should develop photochemistry and photoelectricity much more than it is at present. We should also

use the energy of sun rays both for the raising of low temperatures and for high temperatures, concentrating the light. The energy store of the soil should be not only studied but also controlled, using the great difference in wave-length between the rays of the sun and the radiation of the earth.

3. Physics could not account for the lack of interest in the study of thermo-electricity by its restriction to metals only. As a direct method of deriving electrical energy from thermal sources the thermo-electrical phenomena have to be studied far more closely.

4. New methods of heating buildings are neglected. The idea of using a kind of refrigerator as a heating system propounded by Lord Kelvin could have far more successful application now that the efficiency of our *centrals* has been raised from 15 to 30 and more per cent. Buildings with few outside walls and with a majority of rooms inside without windows might be discussed and have some application.

5. The problem of illumination. Our windows are a very unfortunate method of using the light coming from beyond the earth. We chiefly use the diffused light reflected by the house opposite, which is far from satisfactory. The brilliant reflector signs used in advertising, which are illuminated by light falling from above, show clearly what we lose in our living rooms. We still use glass in our windows and electric lamps which cut off the highly important ultra-violet rays.

6. Powerful beams of high-speed electrons or protons and concentrated electro-magnetic waves could find considerable application within the chemical and electrical industries.

7. The limiting stresses which a physical body can stand were found to be much in excess of the limits actually reached. For instance, we are able to state that an electrical breakdown could be prevented up to a field of over one hundred million volts per centimeter, while we still use a field of forty thousand volts. We have also increased the mechanical strength of crystals many hundreds of times. We have succeeded in discovering substances with an electric constant of over 20,000, while no more than ten are used. An extensive field of investigations is awaiting exploration in order to make the results available to technique.

8. The sensitiveness of the methods developed by physics and chemistry is very striking. We can detect a single electron and proton, and less than one hundred photons of ultra-violet and even visible light. X-Rays and electron rays analysis reveal the finest details of structure. Wireless waves can be detected after they have travelled a hundred

thousand miles. Why have we not adapted these methods to use in everyday life?

There are innumerable other such problems. I am convinced physicists are wrong in neglecting them. Not only would their investigations be of practical use, but they would lead to the development of new problems, would lay bare new features of phenomena known to us only under one aspect. Thus set to work, our interest would lead to a further theory and thence to further experiments, all regarded from one aspect supplied by its origin. New light would be thrown on the old problems and new points of view could be expected as the result of an independent course of research.

We are glad that in our own country we have removed all obstacles to an undisturbed development of science closely bound up with the building of a new future. We have some two thousand physicists. We hope to have the co-operation of millions of workers who are enthusiastic about improvement in their industry and about learning. We do not pursue the policy of keeping the population from science by giving them alcohol, by keeping them 75 per cent. illiterate, by working them so hard that they have no reserve force, as was the case in pre-war days. The more we proceed with improvements in the standard of living, in shortening the hours of work, in increasing the interest in science and art, the more real will become the co-operation of millions of workers in science and technique. By building up the industry of to-day science will simultaneously be working on the great problems of the future. We do not have to fear any resistance from the contradictory inter-play of private interest. All means leading to a higher culture, to better technique, to new knowledge, will be used in order to create a life free from all the burdens of sadness and injustice, borne to-day by the majority of mankind. Science could have no nobler task than that of co-operation in this work.

RELATIONS of SCIENCE, TECHNOLOGY, AND ECONOMICS UNDER CAPITALISM AND IN THE SOVIET UNION.

By M. RUBINSTEIN.

RELATIONS OF SCIENCE, TECHNOLOGY, AND ECONOMICS UNDER CAPITALISM, AND IN THE SOVIET UNION.

Summary of the Report of Prof. M. RUBINSTEIN.

The relations between science, technology, and economics under the conditions of capitalist society and under the socialist system that is being built up in the Soviet Union, are distinctly different, and in many respects, diametrically opposite.

The capitalist system of production and social relations is antagonistic by its very nature. Along with its growth and development there goes on the development and growth of the profoundest intrinsic contradictions that are manifested in all branches of human existence without exception. The purpose of this report is to trace the development of these contradictions in the domain of scientific and technical work and to show how these contradictions vanish and fade away under the conditions of the new system of social relations that is now being built up in the Soviet Union.

It would be useless to describe before this audience the colossal achievements of science and technology during the last century. The report refers only to the basic stages of this development, to its most important present results.

The progress of technical development and the triumph of man over the forces of nature is accelerated with each decade that passes. Substantially speaking, for modern science and technology there are no insoluble problems, and it was quite proper for the American Society of Mechanical Engineers to adopt for their 50th jubilee the slogan: "What is not, may be!"

The development of technology in the epoch of capitalism has proceeded upon the basis of great achievements and growth of the practical application of science. The place of art, of empirics, was taken by exact science, by the application of mathematics, of the laws of mechanics, by investigation into the chemical and physical transformations of substances, by penetrating into the essence of the organic processes of the vegetable and animal world.

Each discovery, each step forward in natural science, has opened new possibilities of industrial development, new conquests for technology. The report adduces a number of instances of modern influences of this kind, which are manifested with particular prominence in the domain of chemistry and electro-technics.

Large scale machine production, constituting the fullest and most striking embodiment of the tendencies of technical development, as Marx said, by its very nature " postulates the replacement of human power by the forces of nature, and of the empirical routine methods by conscious application of science." At the same time the most characteristic feature of all these changes is their fluctuating character, a constant state of motion, revolutionary changes in the technical basis of production, as well as in the functions of the workers and in the social combinations of the process of labour.

Yet, while the technical development was determined to the highest degree by the achievements of science, on the other hand even far more important was the reverse effect. The development of science, including such branches of scientific investigation as would seem to be the most abstract, has gone on chiefly under the influence and requirements of technology. The correctness of this proposition may be demonstrated by thousands of examples from all branches of science.

The report adduces a series of characteristic examples of such kind of effect at the present time, when each of the maturing technological requirements of humanity lends an impetus to profound scientific analysis of natural phenomena, demanding an answer from science to a number of cardinal questions.

It is necessary also to observe that extensive scientific research in the domain of science at the present time cannot be carried on by those individual craft methods which prevailed in this respect even in the 19th century. It requires powerful laboratory equipment, intricate, expensive appliances and instruments, experiments upon a semi-factory scale, a considerable staff for systematic study of the immense literature growing up on each subject.

In the overwhelming majority of cases, it requires the collective organization of labour, the sub-division of the work, and the complex forms of co-operation in this work among specialists in various branches of science, and of various qualifications. Even when carried on by a large collective body, the treatment of many scientific-technical problems takes sometimes years, and even tens of years, calling in many cases for tens and hundreds of thousands of systematic experiments, tests, and observations. In other words, *scientific investigation becomes itself a sort of large scale production* organized after the type of industrial plants. And, however great the obstacles raised in this domain by the particularly lingering traditions of mediaevalism, the development of scientific research work in the advanced capitalist countries has followed precisely this course. For instance, the powerful laboratories of the world's leading chemical and electrical trusts (IG, General Electric, Westinghouse, etc.) have not only become centres where a number of highly important technical discoveries and inventions has been worked out, but they have also been instrumental in creating a series of new scientific theories. In those laboratories there is intense activity going on upon the study of questions which would seem to be most abstract and theoretical.

It seems to me, it would be quite futile here to debate the point as to which came first, the fowl or the egg, science or technology.

As is always the case in life and nature, which develop in dialectical manner, the cause becomes here the effect, and the effect, in its turn, the cause. Moreover, this very distinction becomes more and more conventional, vague, and questionable.

A number of discoveries and theories of the close of the 19th and the beginning of the 20th century have fully undermined, and partly overthrown the rigid system of the division of the sciences in classical science.

Einstein has overthrown the traditional notions about gravity, space, and time. The quanta of the theory has dealt the knock-out blow to the old metaphysical notions about power. Radium, work of Cavendish Laboratory, etc., has turned upside down the old views as to immobile and immutable elements. The study of the laws of electro-magnetic phenomena has enabled us to subject to them the most diverse natural phenomena, having turned upside down thousands of former habitual, deep-rooted notions and theories.

The old, immutable boundaries of the sciences are being obliterated, vanishing just as has vanished the Linnæan system, as has vanished the craft specialization of artisan production.

We are witnessing the progressive development of the so-called "contiguous sciences," such as physical chemistry, bio-chemistry and bio-physics, techno-economic disciplines, etc.

We see how each new economic problem, each new requirement of technology, calls for the collective work of a number of sciences for its solution.

We see how, upon the basis of dialectical materialism, all the sciences are showing a tendency to become transformed into a single system of science (yet permitting of sub-division), into the single science about nature and society spoken of by Marx.

Genuine science studies all phenomena in their state of motion, in the antithesis, and in the development which eliminates the contradictions.

And in this new dialectical unity and sub-division of the sciences, technology occupies its place of equality and honour. It is not merely an "applied" science which used to be scorned by the high-priests of "pure" science and of caste exclusiveness. It is the domain in which man shows primarily his active attitude towards nature, in which he *not only explains, but also modifies the world,* at the same time modifying himself, too.

While the development of technology would have been impossible without science, on the other hand, it is only technology, only industrial practice that can give the incontrovertible answer to a number of cardinal theoretical problems.

While to the priests of pure science it seems a profanation that Marx, in the debate of idealism versus materialism, has appealed to . . . alisarine, and other synthetic dyestuffs, to us, the very division of science into " pure " and " impure " seems monstrous metaphysics.

As was written by Marx in his great theses on Feuerbach, " only by practice should man demonstrate the truth, i.e., the reality and force of his thinking, in his world outlook." It is from this angle of vision that we examine the interdependence of theory and practice, of science and technology, of research work and industrial development.

Approaching the subject in this manner, we at once become confronted with the fact that the development of both science and technology is taking place not in super-terrestrial space, not high up in the clouds, not in the walls of laboratories and scientific studies hermetically sealed off from the rest of the world, but in a distinct social environment, under the conditions of a distinct social system.

Technology and the Contradictions of Capitalist Society.

The social system during the last century was capitalism. And one cannot understand anything as regards the development and the interdependence of science and technology if one tries to examine them apart from a scientific analysis of the rise and decay of the capitalist social relations, apart from the scientific analysis that is furnished by the study of Marx.

The social system for one-sixth of the world has now become Socialism. And one cannot understand anything about the future perspectives of science and technology as well as about the perspectives of their interdependance, without the study of the laws of development, of the struggle and growth of the new socialist system of social relations.

Let us first deal with the first part.

Modern science and modern technology are the offspring of capitalism, and since the latter, by its very nature, is an antagonistic system, there is bound to be equal antagonism under capitalism in the forms of the development and interrelations of science and technology. To begin with, what are the problems of technical progress, and of the scientific development catering for this progress, under the conditions of capitalism?

The purpose and the motive power of capitalist production is the derivation of profits. Whatever the priests of pure science say about profanation, we must observe that under the conditions of capitalism, science as well as technology, whether consciously or unconsciously, serve the interests of capitalist profit.

In outlining the development of the first stages of machine production, Marx quotes a remark by John Stuart Mill to the effect that " it is doubtful whether the mechanical inventions so far made have rendered labour easier even for a single human being "; Marx replies to this : " Neither is this the purpose of machinery used in a capitalist manner. In common with all other methods of development in the productivity of labour, their purpose is to cheapen the price of commodities, to shorten that part of the working day which the worker uses for himself, and thus to lengthen the other part of the day which he gives away gratuitously to the capitalist. Machines are a means for the production of surplus value."*

In this remark by Karl Marx is the whole crux of the question. Capitalism, in developing machine production, pursues the purpose not of developing the forces of production, but of increasing the profits. Therefore, capitalism introduces a new machine only when the difference between the price of this machine and the cost of labour-power that it replaces is sufficiently large to secure an average profit and successful competition in the market. Already at the commencement of capitalist development we find a number of cases when inventions or improvements in machinery were either entirely held in abeyance or they were utilized not in the country where they were originated, because labour in that country happened to be so cheap that the adoption of the machine was unprofitable and undesirable to the capitalists. Marx adduces the example of how a stone-crushing machine invented by Englishmen was not adopted in England because the labourers doing that work were paid such a miserable pittance that the introduction of the machine would have rendered stone-crushing more expensive to the capitalists. A large number of other English inventions was first applied in America for the reason that labour was too cheap in England. For scores of years the European Association of Bottle Manufacturers deliberately blocked on the Continent the adoption of the American machine of Owen for the mechanical manufacturing of bottles. Even the famous Diesel motor was for a long time prevented from being put into use owing to the opposition of coal mine owners whose domination it threatened.

The report contains a minute analysis of the basic contradictions of technical progress and mechanization under the conditions of capitalism, which are demonstrated with particular fullness on the question of unemployment.

Unemployment, under capitalism, is the inevitable consequence of technical progress, and in its turn, it checks the further development of technical progress, the introduction of new machines, and the application of new scientific methods in industrial practice.

*" Capital " Volume I, p. 361, Russian Edition, 1920.

These tendencies to check and obstruct technical, and consequently also scientific development, become particularly pronounced in the final monopoly stage of capitalism.

We can demonstrate a thousand examples of how the powerful capitalist monopolies that have monopolized also the motive forces of technical progress (the apparatus of scientific research work, the laboratories, the patents, and the inventors and scientists themselves) are taking advantage of this monopoly, in the first place, to artificially check the technical progress.

A number of bourgeois scientists and economists, attentively studying the surrounding realities, were bound to admit the rapid growth of these tendencies.

Buying out patents, supporting obsolete plants, fixing cartel prices according to manufacturing costs of the worst plants, secrecy in scientific research work, fear of innovations that threaten depreciation of the old capital stock, etc.—such are everyday facts of industrial reality in the epoch of monopoly capitalism.

Under capitalism, the adoption of technical achievements is always considerably below the extent possible under a given level of scientific and technical development.

As a result we find that in the most advanced capitalist countries the utilization of the achievements of modern technology is limited to a relatively small proportion of plants while allowing the continued existence of obsolete plants in which human labour is wantonly squandered. That the real application of technical discoveries lags far behind the already possible development of the forces of production, is attested by a number of bourgeois economists. Glaring examples of this kind were furnished by the Hoover Commission which investigated the question of waste in industry.

According to calculation by "Iron Age," by putting all the industrial plants in the United States upon the level of modern technique, it would be possible to shorten the working day to one-third of the present, while at the same time doubling the output.

Under the conditions of monopoly capitalism, this discrepancy between technical possibilities and their industrial application becomes particularly great.

Naturally, all these facts and tendencies have a most direct bearing upon the development of scientific research work.

To begin with, these tendencies of monopoly capitalism, by hindering the growth of the forces of production, clip the wings of scientific creative activity, technical initiative, and inventiveness. A huge portion of scientific work, the labour of many years, is practically wasted finding no application in industry, in life, in reality.

As we shall presently see, even a greater portion of scientific thought and activity is squandered upon direct destruction, upon wars and preparations for wars.

Even those scientific achievements which are carried into effect are resulting only in worsening the conditions of millions of toilers, hence the latter are bound to treat them with indifference and hostility. As was written by Marx, " Under capitalism, to be a worker engaged in production is not a blessing, but a curse," and therefore " the worker considers the development of the productivity of his own labour as something inimical to him, and he is right."

This creates for scientific activity an atmosphere of isolation from the overwhelming majority of the population in which, naturally, real scientific creative work cannot be developed to its full extent. Such an extent can be created only under the conditions of the utmost sympathy, support, direct participation of the masses feeling that each forward step in science and technology means improvement in their conditions, relief in their labour, their emancipation. But such a situation we have only in the Soviet Union.

All these contradictions are manifested with particular force in times of capitalist crises.

Under the present world crisis of capitalism, the largest hitherto recorded, which has clearly destroyed all the hopes that were entertained for the possibility of a lengthy period of prosperity without crisis, these effects of capitalist economy on the development of science and technology have manifested themselves with quite unprecedented force.

The report alludes to a number of instances of the colossal waste of forces of production during the period of crisis, the deliberate curtailment of production, the direct destruction of foodstuffs and raw materials, machines and implements.

Science in many cases deliberately and systematically places itself in the service of reducing the food stocks of humanity (e.g., eosination and gasification of rye and wheat in Germany) and the supplies of raw materials. The reduced use of the industrial equipment of the basic capitalist countries to $\frac{1}{4}$-$\frac{1}{3}$ of its capacity leads to the losing of all the advantages of mass production, to increased manufacturing costs, to the transformation of all the achievements of modern technique into hindrance for the capitalists and a source of poverty and destitution for millions of toilers.

No wonder that a number of influential representatives of capitalist industry, technology, science, and of the press, are expressing themselves for slackening the " jazz band of modern industry," for the discontinuance of technical rationalization, for " subordinating technique to the dictates of the merchant," and so forth. The report cites a number of utterances of this kind, as well as a number of attempts at carrying out these ideas in practice (e.g., the " pick and shovel plan " that is being carried out by a number of municipalities in America).

All these theories and plans clearly demonstrate how the conditions of modern capitalism have become an obstacle to the development of the forces of production, of science and technology.

The most stupendous kind of waste of the forces of production under the conditions of modern capitalism is presented by the unemployment crisis.

The fact of there being upwards of 15 million unemployed in the summer of 1930 and of upwards of 20-25 million in the summer of 1931, at the very height of the building and agricultural seasons, the exclusion of over a quarter, and in some countries over half of the working class from the process of production, and the sharp reduction in the consuming capacity of 80-100 million people, implies de-qualification, poverty, starvation, and consequently, emaciation, and partly physical destruction of the basis of the forces of production. This wasting of the most essential of the forces of production by far outweighs the results of all the technical changes, of all the achievements in the organization of production. Tens of millions have to starve and be deprived of sheer necessities for the alleged reason of over-production of commodities. At the same time this becomes no longer a temporary or partial situation, but is more and more becoming universal, lasting and constant for a considerable portion of the population. Just as modern capitalism—in some, although more and more frequent cases—burns or dumps into the sea stocks of food because they cannot be sold with a profit, so it is now " burning " labour power upon an unprecedented scale, not in the process of labour and exploitation, but because it cannot exploit these workers with a profit. The American journalist, Chase, calls this situation the " economics of a madhouse," but Marx has already long before Chase demonstrated that this " madhouse " must inevitably become the basis of capitalist economy.

Bearing in mind the deliberate curtailment of the production of raw materials and foodstuffs, the shortage of work for the staff of employees in production (calculated on the basis of one shift per diem) to the extent of 25 per cent. in " good years " of stabilization, and of almost 50 per cent. at the very commencement of the development of the crisis, the unemployment of a quarter to a third of the workers; taking into account the millions of money that are paid to defray the cost of the last war, the expenditures on current " little " wars, and the incalculable expenditures that are being made in the preparations for future wars, we arrive then at the conclusion that modern capitalism does not utilize even one-hundredth part of the capacity and possibilities of production of the present available production apparatus and man-power. Yet, even in countries of powerful capitalist development, this production apparatus is composed of a motley mixture of modern plants with even larger remains of obsolete, backward production units that are artificially supported by monopoly capitalism, this being done on a particularly large scale in the old capitalist countries.

Bearing in mind, further, the artificial frequently forcible retention of the economic backwardness of the colonies, the enforced backwardness and wanton waste of labour in agriculture; the reparations, the tariff walls, and other numerous obstacles and barriers to the development of the forces of production, we see that in reality, the "co-efficient of useful action" of the modern capitalist machine is even still lower.

If the technical achievements already existing in some of the industrial plants were to be extended, at the present level of technical development, to the whole of industry, transport, and agriculture, then this alone would extend by several times the volume of the forces of production. All this, apart from the unquestionable fact that the further development goes on at an ever increasing pace. Emancipated from the brakes of capitalism, it may yield in the shortest historical periods an unheard of progress in economic development.

A further reflection of the crisis of scientific research work is that in the race for retrenchment, there is a constant diminution in the funds granted towards the upkeep of universities, scientific institutes, laboratories, stipends, etc. Unemployment involving tens of millions of workers, does not spare also the scientific workers, engineers, and technicians. The former Chairman of the German Society of Engineers, Prof. Matschos, draws in the Society's Journal a harrowing picture of the effect of the crisis.

"In the higher technical schools (of Germany) there are about 40,000 students of whom 8,000 annually graduate. Among the graduates there is terrific unemployment. On an average, only 20 per cent. secure jobs, 10 per cent. continue studying, 20 per cent. take on any work outside of their profession, and the remainder, about 50 per cent., are left without any occupation. It is no longer a rare sight to see engineers with diplomas sleeping in doss-houses that open their doors at 10 p.m., who do not enjoy a square meal, who consider themselves lucky if they manage to earn a few marks on any odd job, e.g., as dish-washers, cigarette-vendors, hired dancing partners, etc. Charity tries to take care of most acute cases of distress, but it cannot do the most essential thing—to give these specialists jobs. The mental equipment secured at the price of many sacrifices finds no application.

"They dream of quitting the street, but when asked what they have been doing since having obtained the diploma, they can only reply, 'Looking for a job.' The situation is such that the personnel is everywhere being reduced.

"Yet, thousands of young people flock to the universities. Everybody still believes the profession of an engineer to be rich in promise. At the same time, we find that the societies of engineers are warning more and more about the profession being overcrowded beyond all proportion, warning against all expectations, and demanding a rigid selection. What is going to be the outcome of all this? They are now figuring on 15,000 graduates, but we are told that there are going to be 40,000 of them by 1934. Provision is at present made for about 13,000 academic graduates to be employed in 1934, while there are now 30,000 of them unemployed. Can we afford to contemplate such a situation with folded arms? Is it not high time to put a stop to this mass striving after a diploma and higher learning?"—" V. D. I. Nachrichten," 1931.

The organ of the German industrials, " Deutsche Bergwerkszeitung," commenting on this article (April 21st, 1931), gives a " reassuring " reply to the rhetorical question put by Prof. Matschos, pointing out that in a certain city in Western Germany a group of graduates were generously given jobs as . . . tramway workers. However, the newspaper goes on to say (quite reasonably) that " the warning against academic professions would be far more effective if the warners would at the same time mention professions that are not overcrowded and hold out better promises. This is not done because there are no such professions." Quite so! The newspaper notes also the fact that to a graduate technician the lack of employment implies the end of his career, as there is almost no possibility of adaptation to some other kind of work.

Quite identical is the situation in regard to various groups of intellectuals. As a rule, the conditions of scientific workers engaged directly in scientific research are not any better, but rather worse.

The only way out seen by the professor is to close further admission to the higher schools. These facts show how modern capitalism not only blindly destroys the material forces of production in periods of crisis, not only throws millions of workers out of the process of production, but also tries to cut the roots of future scientific and technical development.

Finally, the crisis introduces into the midst of scientific workers a mass of ideological incoherence and confusion. Unable to fathom the causes of the terrific economic concussions, to give a really scientific analysis of the phenomena taking place around them, and to indicate a way out (all this the Marxian method alone can give), the overwhelming portion of them fall into despondency and pessimism, looking for a way out in mysticism, spiritism, religious superstition, etc. Scientific workers are spending more and more of their time in scholastic exercises, in vain and fruitless attempts at reconciling science with a belief in the supernatural; entrapped in the maze of capitalist contradictions, in the anarchy of the capitalist system, their minds vainly seek salvation in the intercession of those transcendental powers.

The most appalling and ignominious part in the effect of capitalism on scientific and technical development is the rôle played by modern science and technique in the preparations for wars.

The report gives an analysis of the causes which prompt the modern capitalist states to prepare for new military collisions, and the basic technical features of future wars.

The report deals minutely with the incessant systematic activity going on in scientific institutions and laboratories on the preparation of new deadly weapons of warfare destined by their very nature for use not only against foreign armies, but also against the entire civil population of the country.

The greatest achievements of synthetic chemistry, aviation, bacteriology, etc., which serve the needs of humanity, are being adapted to the purpose of wholesale destruction eclipsing all the historic examples of barbarism and savagery. Suffice it to quote the following statement by Mr. Winston Churchill on the character of modern warfare :—

"It was not until the dawn of the twentieth century of the Christian era that War actually began to enter its kingdom as the potential destroyer of the human race. The organization of mankind into great States and Empires and the rise of nations to full collective consciousness enabled enterprises of slaughter to be planned and executed upon a scale, with a perseverance, never before imagined. All the noblest virtues of individuals were gathered together to strengthen the destructive capacity of the mass . . . Science unfolded her treasures and her secrets to the desperate demands of men and placed in their hands agencies and apparatus almost decisive in their character."

After reviewing the great battles of the past, he goes on to describe what he believes the future war would look like :—

"All that happened in the four years of the Great War was only a prelude to what was preparing for the fifth year . . . In 1919, thousands of aeroplanes would have shattered their (German) cities. Scores of thousands of cannon would have blasted their front . . . Poison gases of incredible ingenuity, against which only a secret mask . . . was proof, would have stifled all resistance and paralyzed all life on the hostile front . . .

"These projects were put aside unfinished, unexecuted; but their knowledge was preserved; their data, calculations and discoveries were hastily bundled together and docketed 'for future reference' by the War Offices in every country. The campaign of 1919 was never fought; but its ideas go marching along. In every army they are being explored, elaborated, refined under the surface of peace . . . Death stands at attention; obedient, expectant, ready to serve, ready to shear away the peoples en masse; ready, if called on, to pulverize without hope of repair, what is left of civilization. He awaits only the command."—Winston Churchill, "The World Crisis. The Aftermath," London, 1929, pp. 452-455.

After describing the role of chemical science in this respect, and the pseudo-scientific attempts of some scientists to demonstrate the "humanitarianism" of chemical warfare, the report demonstrates how the war policy exercises the strongest effect on the whole character and trend of scientific research work. Thus, capitalism endeavours in a "planned" manner to subordinate science and technique, the apparatus of production, and the whole population to the task of organized wholesale destruction and extermination. In this respect, the contradictions of scientific and technical development are revealed with particular force, scope, and acuteness.

Already the present state of science and technique secures such a gigantic growth of the forces of production as modern capitalism is unable to realize.

Scores of millions of workers are shut out from the process of production; they are eager to work but they cannot find it.

Other scores of millions are engaged in non-productive labour, in serving the incredibly swollen apparatus of trade, advertizing, the gigantic machinery for suppressing the masses, the manufacturing of public opinion, and, lastly, catering to the luxuries and whims of the upper crust of the bourgeoisie.

Hundreds of millions work from morn till night in factories, mines, plantations, burning away their stamina in a few years, turning old at 40; nevertheless, the social productivity of their labour is relatively negligible as the result of capitalist waste.

Hundreds of millions in agriculture are tied to their miserable plots of land, labouring in the sweat of their brow, under conditions which exclude the application of science and modern technique, not always eking out even the most miserable existence.

Lastly, many millions of workers are still spending all their strength to pay for the consequences of the world war of 1914-18, and the costs of preparations for new wars.

Huge reserves of fuel and metals are waiting in the bowels of the earth to be brought up to the surface.

Waterfalls and rivers are waiting to be harnessed by dams, for the streams of water to set turbines and generators in motion, dispensing the vitalizing current of electricity.

Thousands of technical problems, quite realizable with the present state of technique are still held in abeyance.

Already the present state of science and technology permits, with relatively negligible expenditure of labour effort, the subjection of the elements, erection of new cities, the automatising of a number of production processes, the rendering of labour a joy.

Yet modern capitalism cannot make use of all of these possibilities.

Each attempt on the part of capitalism at the development of the forces of production creates ever new antagonism, leads to ever new and more appalling waste, destruction, crises, and wars. Capitalism cannot help it. No scientific forces can alter these laws which govern the rise and decline of the capitalist society, just as they cannot alter the laws of growth and decay of the human organism. And there is but one science which shows a way out—it is the Marxist scientific analysis of social development.

II. THE SOVIET UNION.

The Soviet Union constitutes the first experiment in human history of the application of this scientific analysis and scientific methods for the conscious construction of social relations, for planned guidance of the economic life, for directing the course of cultural, scientific, and technical development. The very existence and the whole course of development of the Soviet Union is thus connected with genuine scientific theory.

This year the Soviet State is in the thirteenth year of its existence. During the current year has been accomplished more than one-half of the great Five Year Plan of socialist reconstruction.

This makes it necessary for scientific analysis to sum up results, to compare the experiences of two systems, to ascertain their respective tendencies of development. This analysis shows:

Firstly, the unquestionable fact that the appalling world economic crisis engulfing with unprecedented force all the capitalist countries without exception, and all the branches of world economy, is halting at the borders of the Soviet Union. Not only does the Soviet Union not experience a crisis, but on the contrary, during the last two years it has shown a tremendous upward trend of economic development.

Secondly, this comparison shows that while the anarchy of capitalist economy throws millions of workers out of employment, the Soviet Union has disposed of the problem of unemployment, annually attracting millions of new workers into industry, and carrying out a great plan of mechanization to obviate the growing shortage of man-power.

Thirdly, this comparison shows that the tempo of economic development in the Soviet Union is many times faster than in all the capitalist countries, including the United States of America, during their best periods of development.

Fourthly, this comparison shows that while the anarchy of capitalist economy increases year by year and no successes of capital concentration, no efforts of scientific prognostication can soften the spasmodic fits of this fever; in the Soviet Union we see the constantly growing and enduring successes of deliberate *planning* of the entire economic life: the quarterly, annual, and the Five Year Plans are being carried out with a margin; work is now proceeding on the drafting of the second Five Year Plan during which this country is to overtake the leading capitalist countries and gain the mastery of the most advanced modern technique.

Fifthly, this comparison shows that while agriculture throughout the world has been suffering from a crisis for many years already, showing its total inadaptability to reorganization upon the basis of modern science and technique; the agriculture of the Soviet Union, for the first time in the history of mankind, is being remodelled into large-scale collective farming with the most advanced technical methods and new social relations.

Sixthly, this comparison shows that while the conditions of modern capitalism are aggravating more and more the antagonism between city and country, between physical and mental labour, the Soviet Union is taking decisive steps along the road of eliminating these ancient antagonisms upon the basis of drawing the millions of the toilers into the wave of cultural evolution, education, and enlightenment.

Lastly, this comparison shows that while the development of the capitalist antagonisms leads to a distinct intensification of the tendency to check the progress of technology and science; in the Soviet Union science and technology are finding an absolutely unlimited arena for development, quite new possibilities of practical application and of decisive effect upon all branches of life.

All these deductions are based upon facts which no objective, really scientific observer can dispute. These facts may be tested by anyone, and the Soviet Government is prepared to afford to any scientific and technical worker all the possibilities for testing and investigating these facts on the spot.

Notably, the report adduces a number of facts relating to the economic construction now developing in the Soviet Union in all branches of industry, transport, and agriculture. This construction, by its scope, is without precedent in history.

A number of statistical data cited in the report from official capitalist sources (the League of Nations, etc.) show the results of this development in comparison with the development of other countries.

These data, which have already been surpassed in actual life, indicate the results of the contest between the two systems more than volumes of arguments. Suffice it to observe that the index of industrial production of 1930 in all the capitalist countries has sunk below the level of 1925, whereas in the U.S.S.R. it has been *tripled.*

Moreover, the planned utilization of the immense natural wealth of the Soviet Union, and of the even greater reserves of enthusiasm, energy, and creative initiative of the masses, are really only beginning to unfold to their full extent. A declining rôle in this unfoldment is now attached to science and technology.

SCIENCE AND TECHNOLOGY IN THE SOVIET UNION.

The Soviet Union has set before itself the task of technically and economically overtaking and outstripping the advanced capitalist countries within the shortest historical period. The teeming millions of our country are at the present time animated by enthusiasm unknown in history for the mastering of modern science and technique, for the gaining of knowledge which would enable them to remodel the whole of life, to subjugate the forces of nature to the collective will of the toilers. This alone shows the colossal importance attached in the Soviet Union to scientific and technological creative activity, to research work, to the spreading of knowledge among the masses. This, however, does not and cannot limit the rôle and tasks of science in the Soviet Union.

The endeavour to overtake the technique of the advanced capitalist countries does not imply that we can content ourselves with merely copying all the aspects of this technique.

Already in the history of the capitalist world we see that, for instance, the United States, having overtaken and outstripped the technique of the old European countries in the last few decades, was forced to raise and solve a number of quite new technical and scientific problems connected with the requirements of mass production, with the gigantic scope of industrialization in that country.

This applies to an incomparably greater extent to the problems which are at present raised and solved by the Soviet Union that is carrying out industrialization upon an entirely new basis and at a pace and on a scale unknown even to the United States.

Here it has neither previous experience nor examples. Already in the very process of this work it has to solve scientific and technical problems which have not yet been solved anywhere at all.

As a case in point, let us take the domain of agriculture.

Already last year the average annual working of tractors in the United States was 400-600 hours, whereas in the Soviet Union it was no less than 2,500 hours. The Soviet Union already now has thousands of mechanized grain farms surpassing all the records of the United States. In the current year the Soviet Union organizes cattle rearing ranches on a scale unprecedented in the world. It sets before itself the problem of mechanizing all the processes of agriculture in grain growing, commercial crops, gardening, etc. It carries out planned, scientifically thought out specialization of agriculture over vast territories, each of which is equal to the big European countries by its area.

All these tasks call for the creation of new types of machines and implements, for the working out of new forms of connection between the motor and the hitching appliances, for new forms of labour organization, plant selection, etc.

Thus, the technical reconstruction of agriculture involves thousands of new problems in economics, agronomy, chemistry, physics, botany, zoology, energetics, machine construction.

The solving of these problems is unthinkable without the unfolding of scientific research work upon a gigantic scale. And along with the utilization of all the achievements of science and technique of the advanced capitalist countries, utilization in many cases far more complete and effective than in those very countries, the economic practice of the U.S.S.R. already now demands from science and agricultural technique a reply to a number of questions which have not yet been solved, the blazing of new trails, new discoveries and inventions, new scientific theories.

The same relates in equal measure to the problems of electrification of the Soviet Union and to a number of other problems relating to economic and cultural construction.

The completion of the Five Year Plan by next year (i.e., in four years) confronts the Soviet Union with the problem of working out a new, second Five Year Plan. This plan, accompanied by the gigantic quantitative growth of economy, should also afford the most profound qualitative readjustment of the technical basis of the national economy. It stands to reason that the deciding rôle in the elaboration and execution of this plan should belong to science and technology outlining the course of future development.

What is the scientific-technical apparatus possessed by the Soviet Union for this purpose? What are the dynamics of its development, its organizational structure, its relations with other organs of the Soviet State?

The legacy inherited from tzarist Russia in this domain is even more miserable than it is in the domain of industry. Pre-revolutionary Russia had individual great scientists—mathematicians, physicists, chemists, biologists.

They gave a number of important discoveries and inventions, a number of profound, scientific theories, but all those theories and discoveries were in the overwhelming majority utilized only abroad, since neither the feeble industry nor the general atmosphere of the tzarist autocracy—that "prison of nations"—allowed the development and utilization of those discoveries in practice.

Suffice it to observe that in pre-revolutionary Russia there was not really a single scientific research institute worthy of the name. The whole of the scientific activity was concentrated in a few poorly equipped university laboratories that were detached from industry and completely isolated from the masses of the people. In order to furnish an idea of the growth of the network of scientific research organizations under Soviet rule, suffice it to mention that in industry alone there were:—

In 1928.—24 scientific research institutes with 8 branches.

In 1930.—72 scientific research institutes with 83 branches. (Included among these are such gigantic institutions as the Thermo-Technical Institute, the Physico-Technical Institute, etc., which have no equal in Europe).

Agriculture in the current year was served by 47 Institutes, transport—by 10, popular education—by 44, public health—by 34, and so on. The total number of scientific research institutes in the beginning of 1929 was 789.

The number of factory laboratories runs now into thousands. The scientific staffs of industrial institutes (exclusive of factory laboratories, as well as of administrative and service personnel) have reached the number of 11,000. In 1931 there were about 40,000 workers engaged exclusively in scientific research work in this country.

The financing of the network of scientific research institutions in industry alone (again exclusive of the factory laboratories) has reached the amount of about 250,000,000 roubles, as against 12,000,000 in 1925-26 and 58,000,000 in 1928-29.

These fragmentary data testify to a tremendous constant growth year by year.

Nevertheless, even this growth is quite inadequate to satisfy the evergrowing requirements.

The Soviet Government is taking a series of measures to accelerate further the pace of this growth, of the unfolding of the network of scientific research institutions, of the training of the necessary staffs.

The enrolment of students in the universities and technical colleges, which numbered less than 100,000 in 1929, grew into 157,000 in 1931, and has to be further raised to 230,000 for 1932.

Already in 1931 should be achieved the doubling of the number of our engineering and technical personnel, and the full completion of the Five-Year Plan in this respect. Enrolment in the technical schools under the 1932 plan calls for the admission of 420,000 students, of 350,000 students to the workers' faculties (as compared with 166,000 in 1931), and of 1,000,000 pupils in the factory apprenticeship schools as compared with 700,000 in 1931. The proportion of graduates of the workers' faculties in the higher schools should reach 75-80 per cent. This at a time when, according to official German data, among the students of all the higher schools in Germany there are only 2-3 per cent. of proletarian descent, and even in the intermediate schools in Prussia only 5.4 per cent. of the boys and 3.4 per cent. of the girls are of proletarian descent. A bourgeois journal, commenting on these data, observes that " the privilege of higher education is exceedingly rarely won by the sons of the workers. Even if a young worker should pass the examination for matriculation, he would have to work to earn his living. Of the 1,110 lucky ones who got a stipend in 1928 there were only 12 per cent. workers."

In the Soviet Union all the students are assured stipends and board. There is a steady increase in the numbers of proletarian students taking up scientific work upon graduating from the higher school.

Prospective plans for 1932 provide for a 40 per cent. increase in the total number of scientific workers.

One of the most essential features of the organization of scientific research work in the Soviet Union is the principle of *planning*.

At one time there were debates as to whether it was generally possible to plan scientific activity; those debates are now substantially concluded. The socialist plan, which has so brilliantly demonstrated its advantages in the guiding of economy, has been unanimously recognized as the leading principle in the domain of scientific work.

The whole network of research activity in industry is working in conformity with a single summary plan worked out by the

Scientific Research Sector of the Supreme Council of National Economy with the assistance of the Institutes and of prominent workers in various branches of science. The same thing happens in agriculture, transport, and other branches.

In place of isolated individuals whose character and atmosphere of activity is really in the nature of petty craft: in place of the isolated scientific research organs of capitalism that are directly or indirectly subordinated to financial capital, we have here a planned, organized network of scientific research bodies united by the common task of raising the forces of production upon a socialist basis. Recently a new step was taken in the Soviet Union for the planning of the whole of the scientific research work of the country at large. The first Scientific Research Planning Conference, which was attended by over a thousand delegates from scientific organizations in all branches of science and technology, investigated the most essential problems confronting the research workers, outlined the methodology of planning in this domain, appealed to all scientists and scientific workers to join in the working out of this plan. The Conference went on amid tremendous enthusiasm and has demonstrated what inexhaustible reserves of thought and creative activity may become available by doing away with unplanned wastefulness in the domain of scientific work.

The decisions of that Conference may serve to scientific and technical workers of the capitalist countries as an example of the possibilities opened by the Soviet system to scientific thought. For instance, let us allude to the decision to impose the obligation upon all planning and operative economic organs to include in their industrial reconstruction plans, as an organic part thereof, the realization of the achievements of the scientific research institutes furnishing them with the necessary finances and material means.

Or the decision to oblige the economic organization to set apart and attach to the institutes the necessary number of industrial plants to be transformed into experimental works for carrying out the achievements of the new technique. Or the decision to oblige all newly building large industrial enterprises to provide for the installation of factory laboratories as an inseparable part of a given enterprise, or the awarding of premiums to enterprises adopting the advanced technique and fixing legal and material responsibility for delay in the realization of scientific achievements. No less important are the decisions concerning the publication of popular accounts by the scientific institutes on their activities, systematic travelling scholarships for practical industrial workers to take up temporary work in the scientific institutes, the inclusion of directors of scientific institutes upon managing boards of the respective trusts, the widest attraction of the trade unions to render assistance to the scientific institutes and to make propaganda for scientific and technical achievements.

Or let us take the decisions of such types as the inclusion of collective testing of important inventions and improvements in the general plan of the scientifico-technical work of all the branches of industry, transport, and agriculture; the working out of special tasks for inventors by factories and by branches of industry; the submission of plans and achievements of the Academy of Sciences, of the scientific institutes and laboratories, for wide discussion by workers interested in inventions, and so on and so forth.

In no capitalist country would it be possible to achieve anything resembling the measures of this kind. They are incompatible with the very nature of capitalism, they are possible only when science and technology become connnected with the process of the great socialist construction, when the scientific workers, in organized and planned fashion, direct their efforts to the carrying out of the " social order " of the large masses of the toilers—to raise to the highest level the whole technique and economy of the great country that is building socialism.

In this connection it is necessary to observe that even more important than the planning of scientific research work is the direct *organizational connection of science and technology with the large masses of the working class.*

This connection is now beginning to be realized in the Soviet Union upon an entirely unprecedented scale. The struggle for the mastery of science and technique embraces already, not scores and hundreds of thousands, but millions of workers.

This opens up such reserves of energy, initiative, inventiveness, that could not even be dreamed of a short time ago. In each factory, Soviet farm, higher school, special organizations are formed for the mastery of technique, inventor circles, and vast activity is carried on for the spreading of scientific and technical knowledge. During dinner intervals, and in their leisure hours, the great masses of the workers are eagerly and stubbornly studying, attentively watching the possibilities of improvement in their particular line of industry, preparing themselves for admission to technical schools and colleges, enthusiastically welcoming prominent scientists reporting to them on their discoveries and researches. There is only a lack of men and time to satisfy this thirst for culture, knowledge, science, which has arisen even among the most backward strata among the working class. Thus we see the truth of the prognostication made by Engels when he wrote that " the society emancipated from the shackles of capitalist production, bringing forth a new generation of thoroughly developed producers who understand the scientific foundations of the whole of the industrial process and who study practically, each one in his branch, the whole series of branches of production from beginning to end, will be able to create a new force of production." ("Anti-Düring.")

In this manner the antagonism between physical and mental labour begins to be eliminated. Already now, at the very beginning of this development, the struggle of the masses for the mastery of science and technique, is performing miracles. Let us refer, for instance, to the domain of *workers' inventions.* The number of suggestions and inventions by workers has increased a hundredfold during the past year. Frequently one finds factories receiving thousands of suggestions from the workers in the course of the year. Among other things, the struggle of the masses for the mastery of technique reveals itself in the quite novel ways of *organic combination of the planned activity of the scientific research institutes with the mass inventive activity of the workers,* while the latter, in its turn, is connected with an even more powerful movement of the millions—socialist competition and shock work.

Mass inventive activity of the workers is becoming one of the highest forms of socialist competition, one of the most important and most promising stages of its development.

A brilliant example of the first manifestations of this tendency is furnished by the events of recent months in the Donetz Coal Basin.

When the mechanization of the Donetz Basin was taken up as a *political task,* when the carrying out of mechanization became the business of the large masses of the mine workers, the Donetz Basin saw the steady rise in the wave of technical initiative on the part of the workers and the engineering and technical forces. The start was made. And in the recent months there was something like a steady stream of inventions, suggestions, rationalization proposals, all tending to bring about a conveyor flow of coal brought up from the mine, in other words, to bring about a profound technical revolution in the methods of coal mining.

The idea of continuous coal mining originated in the Donetz Basin mines at the end of 1930, when the methods of Kartashev, Kasaurov, Filimonov, and Liebhardt were put forth. This was followed by a steady flow of invention and improvement proposals made by scores of other workers. The proposals are now pouring in from nearly every mechanized pit. Many of these proposals are not even particularly novel. Yet, while analogous ideas were held in abeyance for years in the past, at the present time, combining with the wave of socialistic competition, with the general mighty enthusiasm of the workers, they are bringing about a revolution in the methods of production, foreshadowing in many cases the possibility of not only overtaking, but also outstripping foreign technique in the very near future.

The wave of inventiveness in the Donetz Basin presents an exceedingly telling example of the boundless possibilities harboured in the struggle for the new technique and for industrial improvements rendered possible by arousing the initiative and the spirit of emulation among the masses of the workers.

Lately we saw even more interesting phenomena in this domain. No sooner did the news spread about the imminent underground revolution in the Donetz Basin, no sooner were the general features of the methods of Kartashev, Kasaurov, Filimonov, and Liebhardt made known, when from all parts of the Soviet Union, thousands of kilometers away from the Donetz Basin, in the Siberian mines, in the Urals, in the Kuznetsk Basin, there surged up a similar wave of the inventing initiative. Thus, in the Cheliabinsk coal basin the workers launched the remarkably expressive slogan: "*The Cheliabinsk Pits shall have their own Kartashevs!*" And this slogan did not remain an empty sound. The Cheliabinsk pits did get their own Kartashevs. This slogan was taken up by the large masses of the workers, by engineers and technicians, by scientific research workers. The present slogans are as follows:—

Each factory, mine, Soviet farm, each scientific research institute and laboratory should have their own inventors. Each shock worker, having mastered the technique, may and should become an inventor, a rationalizer, contributing his mite to the improvement of production processes, to the development of technique, and consequently, to the development of science.

In this connection we may refer to another domain in which we see quite similar progress, namely, the *study of the natural resources of the country.* In all international statistical reference books you will find data about the reserves of petroleum, coal, ores, and other mineral wealth upon the territory of the Soviet Union. These data do not reflect a hundredth part of the real resources. Already the discoveries of the last few years have increased the old data tenfold.

Each expedition of the Academy of Sciences and of the geological exploration institutes to Siberia, Central Asia, Kasakstan, Caucasus, etc., reveals new deposits of wealth. *The country is being newly discovered,* in the literal sense of the term. Now this work, besides scientists and special institutes, attracts thousands of voluntary workers among the local population—school teachers, collective farmers, young people. In the most outlying parts of the country there are being formed circles and groups which study the local nature, and after mastering the rudiments of the technique of geological exploration, are enthusiastic in this work of exploring the underground wealth, not for the sake of personal gain, but to assist in the building of socialism. And this movement of the masses, fertilized by scientific thought and modern technique, yields the most unexpected discoveries resulting at times in the total transformation of the economic perspectives of entire districts and regions.

All this promises to give a new mighty stimulus to the "incessant, ever more rapid development of the process of production" prophesied by Engels as the result of shaking off the chains of capitalism.

This development of the force of production postulates similar incessant and ever more rapid development of science.

This outlook is no longer of the distant future, no longer a vague and nebulous aim. It is the very reality in which we are living, working, building. It is the beginning of the new historic stage into which we have just entered.

This outlook is bound to fascinate every honest specialist who loves his work, every scientist and research worker, just as it does the masses of the proletariat in this country.

Thus, the German Professor Bonn was forced to admit, in his book on the United States, that in the U.S.S.R. " the golden age of science and technology has come " and that this fact is of tremendous international importance. Lenin wrote once to the great American electro-technical expert Steinmetz:—

" You, as a representative of electrical technique of one of the most advanced countries in technical development, have become convinced of the necessity and inevitability of replacing capitalism by a new order of society which will establish the planning regulation of economy and will secure the welfare of the whole mass of the people upon the basis of electrification of entire countries.

" In all countries throughout the world there is growing—slower than it might be desired, yet relentlessly and steadily—the number of representatives of science, technique, art, who become convinced of the necessity of substituting for capitalism a different social-economic order, and who, unscared by the ' tremendous difficulties ' of the struggle of Soviet Russia against the whole of the capitalist world, but rather attracted by them, are realizing the inevitability of the struggle and the necessity to take part in it, helping ' the new to overcome the old.' "

Tens of thousands of scientific workers, united in collective bodies and carrying on their work on definite plans, organically associated with the proletariat, constantly drawing re-inforcements from its ranks, blazing new paths for science and technique jointly with the millions of worker inventors and rationalizers, are not only helping to overcome the old handicaps, but also to build up their country anew.

This evolution of progress in U.S.S.R. upon the background of unprecedented crisis of world capitalism is becoming ever more clearly realized by numerous representatives of the bourgeois intelligentsia, by prominent scientists and technicians who cannot shut their eyes to the real facts.

Among numerous statements of this kind, let us refer, for instance, to the remarks by the German economist Bonn on the significance of the American crisis and of the economic construction of the Soviets.

Professor Bonn writes: " The Olympus was wrecked by an earthquake. When the crumbling walls of the temple destroyed the roofs of the huts, and the dying gods, instead of giving protection, deal destruction around them, then the believers are seized, not with regret that the gods too are mortal, but with bitter doubt and blind hatred. What is the sense in worshipping such gods any longer?

" Millions of unemployed, hundreds of thousands of ruined lives, suffering in America under the blows of the crisis: they no longer grumble against individual economic leaders who failed to prevent the crisis, they are beginning to doubt the very system which has made the crisis possible.

" Capitalism and the capitalist economic system hitherto appeared to the average American to be the reasonable form of existence. These forces had built up the greatness of his country in the past, and afforded the opportunities of existence to his predecessors. He expected from them the possibilities of a reasonable existence along the same road.

" This the system can no longer yield. And in thousands of hearts and brains the question arises: has the capitalist system any right at all to exist, if in one of the richest countries in the world it cannot bring about an order of society securing to a relatively sparse, industrious and capable population, an existence that is consistent with the requirements, and with the development of modern technique, without periodically throwing millions of people out of work and damning them to destitution and to the aid of soup kitchens and doss-houses?

" The sense and significance of the American crisis consists in the fact that now not only the present possessing class in America or the ruling class, but the whole of the capitalist system as such is taken under a question mark."*

Professor Bonn observes a profound change in the moods of the intelligentsia, especially of the technical intelligentsia, under the great ideological effect of the Russian revolution, of the very fact of the existence of the Soviet Union. He writes: " Before the Bolshevik revolution it was always possible to object to advocates of socialism that their system was not only wrong, but even if true, it was unrealizable. Now one can no longer brush aside the socialist system as unrealizable. It does exist, and because it exists by the side of the capitalist system, it calls for comparisons." Professor Bonn draws this comparison from the standpoint of the technical intelligentsia of America: " Russian bolshevism implies rigid planning of economy under which the engineer, upon a vacant spot, erects gigantic enterprises with all the means of modern technique. The Americans picture it to themselves as a system which builds up skyscrapers on the prairie at even a quicker pace than it was done in America by private enterprise. This appears to them to be a grand experiment of directing all efforts to the building of a desirable world in place of the old one. The heart of the American engineer on hearing about the possibilities of activity in Russia, beats stronger and faster; because in his own country he cannot think

*Prof. Bonn in " Neue Rundschau," February, 1931.

of erecting greater technical structures than in the past without reducing the profit possibilities.

"The strata of intelligentsia that have gone through the collapse of the American prosperity with its terrible aftermath, are looking in amazement upon the Five-Year Plan which, in their eyes, points the way towards determining the economic fate by a firm hand will . . .

"There is a peculiar charm to the American world emanating from Russia. If the Five Year Plan will be carried out in reality, it will lead many people to the idea that the Russians, who not so very long ago used to be considered as emotional, gifted barbarians, capable of writing the novels of Dostoyevsky or the operas of Tchaikovsky, have now overtaken the Americans in the domain of technique, while in regard to conscious social guidance of society, as demonstrated by their success, they have surpassed the Americans.

"Should the capitalist system fail to draw the millions of unemployed into the industrial process again, the psychological effect of this development will be exceedingly far-reaching."*

Thus, upon the basis of socialist relations in society, overcoming thousands of difficulties and obstacles, combating the numerous survivals of the old, the routine and prejudices of individualism, the Soviet Union is working out the new relations between science, technology, and economics.

It is for this very reason that science in this country, descending from the metaphysical spaces above the clouds, joins in the great problems of socialist reconstruction. It is granted quite unlimited possibilities of development and becomes the leading principle of the whole progress of further construction. While modifying the whole of life, it modifies also itself, starting with the grand remodelling of all the scientific disciplines upon the basis of new methods, of a new monism of all the branches of science. It does not isolate itself from the masses of the workers like a priestly caste; it does not become a hostile force that carries new hardships and privations to the millions of the workers as the involuntary results of its achievements; but on the contrary, it draws ever closer to these masses, steadily obtaining reinforcements from their ranks, and organically joining with the masses in the struggle for common aims and purposes. In this way, it acquires entirely new forces, and opens entirely unprecedented perspectives. The prognosis of Marx and Engels rises more and more clearly, that of the passing of humanity from the reign of necessity into the reign of liberty, where not the machine nor the product governs the man, but the man governs the machine and the product. There is still a difficult road ahead, it will still require a good deal of struggle and many sacrifices, but there is no other way, and overcoming all the obstacles and difficulties, the human race will enter into this world of free and joyous labour by the aid of the subdued forces of nature and of its steel slaves—machines.

*Ibid.

THE "PHYSICAL" AND "BIOLOGICAL" IN THE PROCESS OF ORGANIC EVOLUTION.

By B. ZAVADOVSKY.

THE "PHYSICAL" AND "BIOLOGICAL" IN THE PROCESS OF ORGANIC EVOLUTION.

By B. ZAVADOVSKY.

SUMMARY.

The question of the relationship of the physical and biological sciences, included in the programme of the present Congress, is part of the general problem of the relationship of different systems of world outlook in the solution of the present tasks of natural science. The solution of this problem has repeatedly changed its forms, according to the particular conditions of the working experience of mankind, the condition of its material forces of production, and its socio-economic productive relations, which have been constantly changing in the course of human history. For this reason the extent of my subject does not permit me to reply to the question propounded in all its quantitative volume, and suggests the decision to deal with a few points of principle which lead to the solution of the problem as a whole, examining the question of the relationship of the physical and biological sciences in the solution of some single theoretical problem of biology. As such a problem, I will take the theory of organic evolution—the more because in analysing this problem it will be possible to make some observations on other questions in the Congress programme: the relationship of theory and practice in scientific work, and the role of the historical method in the solution of problems of natural science.

With all the variety of existing opinion in bourgeois science on the question of the relationship of the physical and biological sciences, it is possible to distinguish among them two basic and mutually exclusive tendencies: either (1) attempts to identify the two, reducing biological phenomena to laws of a physical character, or (2) a sharp contrasting of the biological to the physical, as two opposite entities. In the latter case, by "physical" is understood the material forces of inorganic nature, or "mechano-physiological" factors at work inside the organism and reducible in the final analysis to the same mechanical laws of molecular motion: while by "biological" is understood some vital forces of a non-material and non-spatial character, which "are neither the result nor the combination of physical and chemical—i.e., in the final analysis of mechanical phenomena."

In spite of the multiformity and variety of contradictory forces and interests functioning in capitalist conditions of production, it is nevertheless not difficult to establish the predominance of the views of mechanical materialism in the period when capitalism was in its prime as an economic system, and when material culture was rapidly growing as a result of the successes of science and technique—at the end of the xviii. and during the xix. century: and the rebirth of idealistic, vitalistic and even mystical moods, in the measure of the growth of economic contradictions and the sharpening of the class struggle in bourgeois society.

These tendencies acquire special force in the present period of general decline and decay of capitalism, which find their expression also in those contradictions which are delaying the further successful development of natural science and technique under bourgeoisie methods of production: and when, on the other hand, the growth of scientific knowledge reveals the impossibility of reducing all the complex phenomena of nature to a single formula of physical or mechanical laws. These tendencies characterise the general disillusionment of bourgeois society in the possibilities of material culture, and the recognition of the hopelessness of solving the scientific problems which have matured while remaining within the framework of the capitalist system (cf. the report of the Prussian Minister, Dr. Becker, "Educational questions in the period of the crisis of material culture").

This struggle of two systems of world-outlook finds its natural reflection in the existing currents of evolutionist doctrine, which strive to solve the same problem of the relationship of the "physical" and "biological" as factors of organic evolution. In this case the "physical" is frequently identified with the surrounding "external" conditions, and the biological with the "internal" autonomous vital forces, "entelechies" or "dominants," immanent and inherent in life as such, in contrast to the material, "physical" laws of nature.

The principal characteristic of this struggle, and of the ensuing fluctuation in the relationship of the physical and biological sciences throughout the whole history of natural science, is the uncritical use of the conceptions of "physical" and "biological," of "external" and "internal," and the absence of any form of principles of philosophic method, which distinguish the overwhelming majority of the representatives of empirical science.

Thus, within the framework of the conception "biological" itself, there is not always a sufficiently sharp distinction drawn between the idea of "biophysiological," as a factor which determines chiefly the processes of individual development, of metabolism and the regulation of the activity of organisms (although this "biophysiological" also inevitably includes the historical element also), and the idea of "biohistorical," as a factor in the formation of species and phylogenesis.

There are also not infrequent tendencies to include in the "biological" also phenomena in the social history of mankind, since human society is regarded as a simple mechanical sum of human biological species.

On the other hand, there are frequent identifications of the "external" in the process of organic evolution with the physical, and of the internal with the "biological"—forgetting that the biological includes physical, chemical and physicochemical factors as the moment and necessary condition for its realisation, while the "external" in regard to a particular organism in its turn is composed not only of the physical conditions of inorganic nature, but also of the biological surroundings of other organisms, in the midst of and in interaction with which the life of the species proceeds. As for man, the "external" consists first of all of socio-economic productive relations and the condition of material productive forces, by which the socio-historical process is determined.

It is extremely characteristic of the endless contradiction in which modern empirical natural science has become involved that none of the theories of evolution existing in bourgeois science is able to maintain itself in the positions it selects for itself, but slides into the very positions which it was called upon to refute.

Thus neo-Lamarckianism, originally basing its objections to Darwinism on the alleged "unscientific" character of the idea of chance upon which Darwin based his theory of selection, and his attempts to provide a materialist justification for the facts of variability of organisms and their adaptations (and consequently for the whole process of formation of species) in the "direct equilibration" of the organism in relation to the influences of the outside physical surroundings, transfers the problem of adaptation, from the sphere of the rational study of the complex relationships arising between the organism and the external milieu of its existence, into the organism itself. Thus it arrives at the vitalistic and teleological conceptions of immanent vital forces which determine the course and direction of the process of evolution.

Thus, again, the "mechanico-physiological" theory of Nageli, or Berg's theory of Nomogenesis, in spite of all the efforts of the authors to prove the strictly scientific and materialist content of their constructions, arrive at essentially vitalistic ideas of the "principle of perfection," or to the idea of adaptation as the "primary physico-chemical quality of the living matter"—ideas which cannot deceive anyone by their outwardly materialist phraseology.

Thus, again, frankly vitalistic theories, which raised the banner of struggle against the vulgar materialist conceptions of mechanism, strive to find a road to the knowledge of the nature of biological phenomena through non-cognisable and non-material forces, contrasted to the physical world. On the other hand, they are obliged to advocate "practical vitalism"—i.e., the advantage of those same mechanistic methods of research in the practical activity of the research worker. Thereby they pass to the positions of vulgar mechanism in all spheres of direct cognitive action, condemning thereby their vital forces and entelechies to the barren rôle of a bashful screen for our ignorance.

And thus the geneticist, uncritically developing neo-Darwinist ideas of the independence of the germ-plasm of all the "physical" influences of the external surroundings, objectively arrives at the position of the autonomy of the "biological" from the "physical." Thus he descends to those very ideas of autogenesis maintained by his Lamarckian opponents, or to the conception of evolution as the result of the combinations of eternally existing genes—i.e., in fact to the negation of the very idea of evolution, as a process of the continuous unfolding of new formations in nature.

Finally, the Lamarckian, considering evolution as the result of hereditarily accumulated somatic changes, enters the past of the same mechanistic identification of the "biophysiological" and the "biohistorical," forgetting the qualitative peculiarity which distinguishes the ovum, only containing within itself the potential possibilities of further development into a complex organism, and the developing organism in its realisation. In the last analysis this point of view is once again the negation of the very fact of development as an independent historical process of new formations, representing as it does the ovum as the miniature model, so to speak, of the future form, and reducing the process of development in reality to the functions of growth.

The same insoluble internal contradictions mar the numerous attempts to solve the problem of organic evolution by means of an eclectic reconciliation of the Darwinist position with Lamarckian ideas (Haeckel, Plate, Darwin himself, who accepted—albeit with a grimace—the Lamarckian idea of inheritance of acquired characteristics, and many others), since the logical conclusion from the Lamarckian idea of the direct adaptive group variation of the organism, in reply to one and the same influence of external surroundings, is the uselessness and impotence of selection, as a factor in the formation of species—i.e., the negation of Darwinism.

A striking example of the helplessness with which the most outstanding representatives of bourgeois science hesitate between the mechanistic "reduction" of biological processes to physical, on the one hand, and recognition of the absolute autonomy of the biological, on the other, is the position of Professor Muller ("The Method of Evolution"). Having first proved with irreproachable lucidity the fact of dependence of germ plasm on the action of the Röntgen ray—i.e., on physical influences of the external surroundings—so strongly denied until recently by the majority of geneticists —and taking his stand in general upon correct Darwinist positions, Muller nevertheless returns, albeit with many reservations, to the at bottom mechanistic proposition to consider the process of variation as the direct result of the influence of Rôntgen rays upon the germ plasm. Thus he reduces the problem of the modification of the gene, as a biological factor of heredity, to the physical moment of the expulsion of an electron from the biological molecule, forgetting thereby the profound qualitative peculiarity of the biological process compared with physical phenomena.

The final outcome of this crisis through which the theory of evolution is passing in the countries of capitalism is the attempt completely to deny the very fact of evolution, or to consider the theory as one of the possible "hypotheses," circulating side by side with the Biblical legend of the creation of the world in six days' labour, or finally the position of frank agnosticism and disillusionment as to the possibility of solving the problem of evolution at the present level of scientific knowledge (Johansen, Batson, and, in the U.S.S.R., Filipchenko).

From the *socio-historical* standpoint, these schools of thought are the result and reflection in the consciousness of the bourgeois scientist of the internal social-economic contradictions which have gripped the countries of capitalism, and express the impossibility of the further normal development of natural science, as of all sciences, in the framework of the capitalist system.

From the *methodological* standpoint, these positions are the result of the contempt displayed up to the present by naturalists, carried away by the empirical successes of their sciences and the growth of their technical application, for the tasks of the philosophic methodological review and mastery of the facts and conclusions studied in their branch of science. To the extent that individual scientists make attempts at such philosophical generalisations, the positions set forth above reflect their inability, in virtue of the class limitations upon their general train of thought, to adopt the only correct philosophic positions of dialectical materialism.

"Naturalists imagine that they are emancipating themselves from philosophy when they ignore or abuse it. But as they cannot stir a step without thought, while for thought logical definitions are necessary; and these definitions they incautiously borrow either from the current theoretical property of so-called educated people, who are dominated by the remnants of long-passed-away philosophic systems, or else from their uncritical and unsystematic reading of all kinds of philosophical works: in the long run they prove after all to be prisoners to philosophy, but, unfortunately, for the most part philosophy of the very worst quality. And so people who are particularly vehement in abusing philosophy become the slaves of the worst vulgarised relics of the worst philosophical systems." (*F. Engels*: "Dialectics of Nature," p. 25.)

There exists also the firm but incorrect impression that the task of science in general is at all costs to reduce the more complex phenomena to the more simple, and that consequently the successes of the biological sciences are possible only in the shape of the reduction of the phenomena of life to more simple physical rules, while the social sciences can build their laws only by relying upon the achievements of biology. In reality we see that, for example, the facts of heredity, which seemed relatively simple in the days of Darwin—when they were treated of in the Lamarckian, man-in-the-street sense of the transmission by heredity of acquired characteristics, an interpretation very attractive by its apparent simplicity—have received their true explanation to-day only in the very complicated formulae of Mendelism and Morganism. Many remarkable physical phenomena were first discovered by biologists, and many laws of their effect upon the living organism were established before their physical nature became known (X-rays, phenomena of animal electricity, etc.). The fundamental laws of development of human society, which make it possible in our time for the population of one-sixth of the globe successfully to surmount difficulties of what would seem an unequal struggle, were discovered by Marx and Engels 20 years before Darwin formulated the fundamental laws of organic evolution.

All this shows that the true task of scientific research is not the violent identification of the biological and the physical, but the ability to discover the qualitatively specific controlling principles which characterise the principal features of every given phenomenon, and to find methods of research appropriate to the phenomenon studied. This is why if, within the framework of the same physical sciences, we learned to understand that water by no means represents a simple mechanical mixture of oxygen and hydrogen, but constitutes a *new quality* in the physical and chemical properties of *water,* all the more do the phenomena of life represent a complex material system, requiring for its study special methods of bio-physiological and bio-historical research. These laws, as for example, the law of natural selection, or the physiological laws operating within an organism, are in some sense no more and no less simple or complex than the physical laws conditioning the movement of the planetary system, or the movement of electrons around the atomic nucleus.

The fundamental consideration to be borne in mind in this problem is the impossibility of a simple, crude identification of these two categories of phenomena, and the futility of attempts to reduce biological laws to physical, just like the attempts of the vitalists to comprehend the phenomena of the world from the standpoint of the universal animation of matter.

Asserting the reality of the world existing objectively outside ourselves, dialectical materialism starts from the conviction, justified by all the practice of human activity, that our consciousness reflects not only the objective reality of the facts directly perceived by our organs of sensation, but also the constant order of the relations connecting these facts one with another: the fact and its ordered relations with the other surrounding facts are considered by dialectics in their indissoluble unity and entirety. This obliges us to accept not only the *facts* of similarity and unity of structure of organisms, but also that sole possible and rational explanation of these facts which lies in the recognition of the unity of their origin and in the historical law of *development*, which interconnects all phenomena in nature with one another. Hence for us evolution is just as unquestionable a fact as the facts directly perceived by us, of the existence of the ape and man separately from each other.

Establishing the fact of development, variation, motion as the basic qualities of matter, and the unity of the fundamental laws of dialectics, binding on all forms of motion of matter (the law of the unity of opposites, the law of negation of the negation, and the law of the passing of quantity into quality and vice versa), materialist dialectics at the same time emphasizes with all its force the extreme multiformity and the specific qualitative distinctions of the various forms of motion of matter, and the laws characteristic of the different stages of development of matter: and consequently the necessity of the existence of special independent sciences studying these different forms of motion.

In this respect the dialectical conception of universal development—proved by Hegel and materialistically refashioned by Marx, Engels, and Lenin—covers the Darwinian theory of organic evolution, which is the concrete expression of the dialectical process applied to the biological form of motion of matter, and at the same time makes it possible to overcome a number of methodological errors and contradictions on these questions accumulated within the limits of bourgeois natural science.

Precisely from this point of view, biological phenomena, historically connected with physical phenomena in inorganic nature, are none the less not only not reducible to physico-chemical or mechanical laws, but within their own limits as biological processes display varied and qualitatively distinct laws. Thereby biological laws do not in the least lose their material quality and cognisability, requiring only in each case methods of research appropriate to the phenomena studied.

The necessary consequence of the above is a conclusion as to the dialectical development of matter by leaps, bound up with qualitative revolutionary changes as a result of the accumulation of quantitative changes, and the idea of the *relative autonomy* of the biological process, advancing not only in circumstances of interaction with the physical conditions of its surroundings, but also as a result of the development of the internal contradictions latent in the biological system itself. By this means are overcome the over-simplified mechanistic attempts to conceive of the biological process of development as the result of only the physical influences of external surroundings, or of similar physical and physico-chemical processes inside the organism itself or its genes, by which means, it is alleged, it is possible to explain the most complex and qualitatively peculiar phenomena of mutatory variation, and thereby the whole process of formation of species. At the same time this standpoint also overcomes the metaphysical opposition of the biological to the physical, as an absolutely autonomous and independent principle, to the extent that this biological is considered in its indissoluble historical connection with physical phenomena (as a higher form of motion, originating out of lower inorganic forms of motion of matter), and also its dynamic connection (metabolism).

At the same time dialectical methodology by no means eliminates the rôle of the external and physical in the process of organic evolution, requiring only a sharp definition of these conceptions in each case, and the recognition of the multiformity of all those forms of connection which exist between organisms and their external surroundings, between the "biological" and the "physical." Thus the physical constitutes the necessary *condition* in the framework of which the biological process takes place, but at the same time it enters as a necessary aspect into the biological process as such. Furthermore, it may be the direct stimulus of mutatory variations in the germ plasm, thus simultaneously being both external and internal in relation to the "biological." Finally, it may serve as the controlling factor which, in the process of natural selection, determines the very course of the evolutionary process, and therefore acts as the creator of biological forms. In this way the "external" is composed not only of the physical conditions of the external surroundings, but also of the biological encirclement by a milieu of other organisms, and also—in the case of the evolution of man—the social-economic relations prevailing within human society.

Differentiating the conception of the biological as an expression of ontogenetic development, on the one hand, and phylogenetic development on the other, materialist dialectics considers phylogenesis as a particular, most complex form of interaction of the "biological" and "physical" (the organism and its surroundings) and of the biological with itself (the biological relationship of organisms). In this conception there are "eliminated," as it were, or retire into the background, both the purely physical laws of the external surroundings, and the "biophysiological" laws of individual development, qualitatively submitting to the new specific laws of historical biology.

Only in virtue of these new relations, regulated by the Darwinian law of the struggle for life and natural selection, do individual inherited variations acquire the force of a factor in the formation of species, and can the most complex phenomena of biological adaptation (such as protective colouring, mimicry, care for the progeny and the other instincts, parasitism, symbiosis, etc.), receive their rational materialist explanation.

At the same time there finally collapse the equally barren attempts to embrace all the complexity and multiformity of the world through a single mathematical formula of the mechanical movement of molecules, or through the vitalistic idea of a single "principle of perfection," in effect representing an attempt to know and explain the world through the inexplicable and the unknowable.

One of the forms of consciously or unconsciously accepted mechanistic views on the nature of things is the attempts mechanistically to transfer biological laws to the sphere of social and historical relations, in which once again is forgotten the fundamental dialectical law of the qualitative peculiarity of the laws appropriate to every form of motion of matter. These attempts, in the shape of so-called "social Darwinism," strive to find in the biological law of the struggle for existence a justification for capitalist competition, racial and class inequality, and war as a factor of "selection." While they reveal with peculiar vividness the class limitations of scientific theory, and the rôle of the bourgeois scientist as the ideologue reflecting the interests of his class, at the same time these theories suffer from the basic methodological defect of failing to understand all the specific conditions, in the shape of social-economic productive relations, which condition the laws of the social-historic process, allotting to biological factors a remotely subordinate importance.

At the same time, even remaining within the framework of biological factors and laws, we cannot but remark the patently arbitrary interpretation of biological facts on the part of bourgeois eugenists, who attempt to consider the social inequality of men as the direct result of biological inequality in their inherited characters. Since, apart from the relativity and class content of the very conception of a "better" and "worse" genetic fund, it is precisely the biologically established facts of the persistence and resisting capacity of hereditary characteristics, in relation to the influences of external surroundings, and not the Lamarckian point of view, which necessarily explain the fact, confirmed by the objective course of history, that, *notwithstanding* unfavourable external conditions—agelong underfeeding, unemployment and other privations connected with poverty—in the ranks of the working class there grow up ever new fighters for a better future for humanity, while the country building Socialism has at once found its own military leaders, its builders of national economy, science and technique, who have been able to provide the best examples of planned work and organisation of national life.

It is quite normal that the industrial bourgeois class, progressive in its day, saw in the consciously formulated positions of materialist radicalism a theoretical support for its struggle against the influence of the Church and the religious-idealistic ideology which served as a support for the conservative forces of feudalism. That is why the materialist nucleus of the Darwinian theory was at first received with approval by the ideologues of the bourgeoisie, as a scientific proof and justification of the principles of free capitalist competition. And it is just as normal that, in the measure of the growth of economic contradictions, we observe in present-day scientific literature of the bourgeois West more and more frequent attempts to revise Darwinism, and to return to patently idealistic and mystical conceptions—up to and including the open persecution of evolution (the monkey trial in America), and the quest in the embraces of the Church and the Bible for the reply to problems of the universe and for the revival of waning faith in the stability of the capitalist system.

All these facts prove the socio-historical and class determinateness of scientific theories.

Reflecting the state of the material forces of production and the socio-economic relations of the particular historical epoch, scientific theories express not only the actual state and level of knowledge attained by science, but also the ideological justification of the economic interests of warring groups and classes. At the same time they represent a guide to action in the hands of the social groups sharing the theory concerned. That is why the proletariat, fighting for the social reconstruction of the whole world and laying the foundations of a new society and a new culture, is faced with the task of critically reviewing the whole of the heritage received by us from bourgeois science, and of overcoming the theoretical structures which, while not following from the true correlation of things, at the same time expose the class features and purposefulness of the social formations which created that science in the past. The necessity of this is dictated not only by the common interest in cognition of the truth of the world surrounding us, but also by the immediate interests of the struggle of the working class for its emancipation from the economic yoke and ideological influence of hostile classes, in the countries of capitalism, and by the practical problems of Socialist construction in all spheres of national economy in the U.S.S.R., organised by the proletariat on the foundations of the scientific study of the laws of development of nature and human society. Herein lies the cause of the profound interest in and attention to scientific theory, to scientific theoretical research, and to history of sciences, which are displayed in the Soviet Union.

The correct definition of the relationship of the biological and physical sciences, and in particular the relationship of the "physical" and "biological" in the biological process—on the one hand of individual development, and on the other of the formation of species and production of new breeds of domestic animals and cultivated plants—becomes of vast significance in the planned solution of the problems of large-scale Socialist agriculture and cattle breeding. These necessitate the overcoming both of the mechanistic and Lamarckian conceptions, widely held among the majority of practical cattle breeders, which seek a solution of the whole problem in artificial physical influence on the organism: and of the autogenetic enthusiasm of the geneticists, who think that the tasks of the Socialist Five Year Plan are covered by the application of the methods of modern genetics and selection, ignoring the rôle and importance of the rest of man's system of social measures based on the influence of the external physical surroundings on the development of the phenotype and the possible emergence of new inherited variations.

Finally, these theoretical conclusions are of no less importance in solving the practical problems arising out of the reorganisation of the whole system of pedagogy, and of the scientific reconstruction of physical culture, sanitation and hygiene of the human body, which also require for their adequate solution on each occasion theory tested by fact and methodically thought out, and relying *inter alia* also upon a correct definition of the relationship of the physical, biological and socio-historical sciences.

Affirming the unity of the universe and the qualitative multiformity of its expression in different forms of motion of matter, it is necessary to renounce both simplified identification and reduction of some sciences to others, as the supporters of the mechanistic and positivist currents in the sphere of natural science strive to do, and sharp demarcation and drawing of absolute watersheds between the physical, biological and socio-historical sciences—which frequently take the form of admitting the existence of the causal determinateness of phenomena only in the sphere of physical science, while proposing to seek in biological science for teleological solutions, and in the sphere of socio-historical phenomena completely abandoning the search for any order and explanation of the course of historical processes at all.

Since the concrete reality of the phenomena we study is in unity and complex interaction with the whole totality of surrounding phenomena, every exhaustive and worth-while piece of research requires the consideration and drawing in of all contiguous branches of science and the particular methods of research which they represent, and at the same time the subordination of all the sciences to the single gnoseology and methodology of dialectical materialism.

The numerous attempts to revise the conceptions of mechanical materialism—unsatisfactory to the modern naturalist, but the sole conceptions with which he was familiar—without falling into the embraces of vitalism, are condemned beforehand to failure so long as the naturalist remains within the bounds of a methodology based on formal logic and of metaphysical searches for the essence of things, as isolated absolutes, irrespective of their connection and interaction with surrounding phenomena, and without taking into account those variations, that motion, which characterises the dialectical development of the whole world.

At the same time these searches bear witness to the fact that modern natural science is undergoing a profound crisis, hindering its further normal development, and that the general level of knowledge attained is ripe for the conscious application of the dialectical method.

All the more is it a matter for regret that the modern naturalist, when studying problems of philosophy and the history of natural science, remains unaware that these problems of overcoming, on the one hand, the most reactionary idealistic and vitalistic currents of thoughts, and on the other the oversimplified mechanistic positions of vulgar materialism, were not only formulated but solved in their basic and characteristic principles more than seventy years ago, in the classic works of the founders of the philosophy of dialectical materialism, Marx and Engels, and in our own times in the profound works of Lenin.

DYNAMIC AND STATISTICAL REGULARITY IN PHYSICS AND BIOLOGY.

By E. COLMAN.

DYNAMIC AND STATISTICAL REGULARITY IN PHYSICS AND BIOLOGY.

By Prof. E. COLMAN.*

The problem of the character of regularity plays a decisive part in the history of philosophy, of the natural and social sciences. And even though dynamic and statistic regularity exhaust all varieties of laws in the material world and in the consciousness that pictures it (since these are dialectical laws) as little as causality exhausts all types of relations, though the entire problem is closely connected with other questions of main philosophical categories, such as necessity and freedom, causality and chance, continuity and discontinuity, etc. in their application to physics and biology, nevertheless the problem itself does form a certain whole and is sufficiently in the centre of daily interests to serve as the theme for a separate discussion.

Without an understanding of regularity from the standpoint of dialectical materialism, physics and biology cannot steer a way through the Scylla of mechanistic fatalism and the Charybdis of indeterminism. But special importance attaches to the question, because as the crisis in capitalism develops, bourgeois world science grows more reactionary, making irresistibly towards unconcealed *fideism.*

If, twenty years ago, bourgeois philosophers, despite the spontaneous materialism of the majority of physicists, saw in radio-active disintegration of matter a proof of the disappearance of matter, and if, further, the theory of relativity has been taken as affirmative evidence of philosophical relativism, subjectivism, the bourgeois physicists of to-day, resting their arguments upon the quanta theory, declare that causality has been overthrown and solemnly place upon the throne thus vacated the *causa finalis* of Aristotles. From the multitude of recent instances we shall select a few: in his article *Die Kausalität in der gegenwärtigen Physik,* published in the February number of *Die Naturwissen schaften,* M. Schlick regrets his earlier errors. He admits that he made all too great concessions to materialism and adopts the viewpoint that past and future cannot be differentiated, that natural phenomena are undetermined and indeterminable, denying causality and determinism.

In his address, published in the January number of *Nature* under the title *Present Status of Theory and Experiment as to Atomic Dèsintigration and Atomic Synthesis*, R. A. Millikan drew a comparison between three stages of theological thought

* The biological material was supplied by Professor A. M. Krinitzky, Vice-Director of the State Institute of Micro-Biology (Moscow).

on the relation between God and the World: first, the conception resting upon the second principle of thermo-dynamics, of a God that was necessary to wind up the slowing down world mechanism; second, the theory based upon the Darwinian theory of evolution, of the identity of God with the world, which it is claimed, represents the philosophical attitude of most great scientists from Leonardo da Vinci to Newton and Einstein, and finally, the return to the medieval theistic theology of an unbroken act of creation, with which Millikan expresses a certain sympathy in the following words: "This has been speculatively suggested many times before, in order to allow the creator to be continually on his job. Here is, perhaps, a little bit of experimental fingerpointing in that direction," and which, it is alleged, finds support in the discovery of cosmic rays.

A. S. Eddington, in *The End of the World from the Standpoint of mathematical Physics* (in the March issue of *Nature*, takes up the position that the world is spatially finite, has a beginning in time and is developing towards a greater and greater lack of organisation. To us Marxist-Leninists it is obvious that this physical theory merely reflects the general tendency in bourgeois ideology, which interprets the approaching and inevitable end of the capitalist system as the approach of anarchy.

Nevertheless, viewed even from the theological, reactionary aspect, such a theory can offer little to the investigator. From the data adduced by Eddington himself:—An original world radius of 1,200 million light-years, a world radius to-day ten times as great, and the fact that the world radius doubles every 1,500 million years, it can be calculated, without much difficulty, that God created the world about 5 milliard years ago. This, it is true, indicates some error in the Bible, but it also contradicts the period of billions of years, accepted by Eddington himself elsewhere, that is necessary for the origin of stars and chemical elements.

In times of crisis the ideological pressure of the ruling class upon scientific creation is exerted more powerfully than at any other time.

In the social sciences the determination of the character of laws, their cognitive power—the scope of their content and their ability to serve scientific predictions—and their limits, is as important as it is in the natural sciences. In the Soviet Union, where the foundations of Socialist economy are nearing completion, where, consequently, the basis for free play of market forces is continually growing narrower, being thrust aside and replaced by socialist planning, questions relating to the validity of statistical methods, to the reliability of predictions, to methods of planning as a whole, are extremely acute and cannot be answered in the absence of a sure methodological basis. The problem of planning, of subjecting elemental forces to planned direction, is the vital point both of the theory and the practice of economy and policy.

How is the problem of dynamic and statistical regularity raised and solved by contemporary philosophical tendencies in the natural sciences? The two extremes are represented by mechanism, which in truth only recognises the rule of dynamic laws, interpreting them in a narrowly mechanical fashion as the spatial translation of particles without quality, and by indeterminism, which disguises its denial of necessity and conditionality by recognising only statistical laws, alleged to be the expression of free, undetermined chance.

The mechanistic conception of regularity that is characteristic of the metaphysical period of the natural sciences in the 19th century, was formulated most clearly by H. Helmholtz in the following words, taken from his Innsbruck address in 1869 *Ueber das Ziel und die Fortschritte der Naturwissenschaft*: "If, however, movement is the primary change underlying all other changes in the world, all elemental forces are forces of movement and the aim of the natural sciences is to find the movements underlying all other changes and to find their motive power, that is, to resolve them into mechanics."

For our purpose to-day, we are less interested in the unscientific identification of movements with mechanical space displacement than in another no less significant methodological error committed by Helmholtz: he does not understand the difference between the general and the particular. Engels, who understood quite as well as Helmholtz that every movement is connected in one way or another with mechanical movement but who, unlike Helmholtz, was a dialectician, did not propose to investigate the general, the average and the indifferent but, on the contrary, the particular feature in which one kind of movement differs from another. We would add that the inability to recognise the particular is closely connected with the mechanistic dismemberment of matter into identical atoms without quality and that the assertion that dynamic laws are alone objective necessarily implies that statistical regularity is recognised as being valid only subjectively (in nature it is alleged not to exist, it merely reflects our ignorance).

In biology the same methodological error underlies the mechanistic conception as in physics: a false understanding of the relations between the general, particular and individual that, in its turn, can be explained by the theory of knowledge which, subjective in its nature, denies the objectivity of quality to the individuals treated by biology, reducing them to physical-chemical processes and finally to mechanical movement. This subjectivism may possess a rationalist character, when the particular is deduced from the general, given *a priori*, or it may be of an empirical character, when the general itself is regarded as a subjective category and is built up of individual parts. Thus the connection between the general and the particular is reduced to one of quantity only. The univocacy of natural phenomena is a necessary premiss to this, for only on this condition can dynamic regularity be applied, and mathematical apparatus be brought into play, without further ado, upon all the complicated processes and forms of the organic world.

The attempts that have been made to mathematicise biology are extremely characteristic, for their inadequacy, their inability to comprehend the multiplicity and diversity of the subject, is at times clearly apparent, particularly in concrete matters. We would take as an example the work of Ronald Ross: *The Prevention of Malaria,* in which a complete system of equations has been elaborated to portray the dynamics of malaria epidemics although the author is finally compelled to admit that it contributes but little to the elucidation of the actual conditions. Mühlens, the German investigator of malaria, is quite right in remarking that the extent and the severity of malaria epidemics are not dependent simply upon the number of flies and parasite-carriers, but also upon numerous other factors, including not only general, hereditary, climatic and seasonal conditions, but also the individual disposition of flies and human beings etc. Ross' purely quantitative calculations cannot, however, reflect all these qualities.

Still more significant is the attempt to portray biological laws of the relations between organic species living together, of conditions of hereditary transmission, in differential equations, an attempt made by Volterra in which he formulates three main mathematical laws of the fluctuations of species living together: (1) The law of the periodic cycle; (2) the law of the conservation of the average; (3) the law of the perturbation of the average. Volterra himself realises that mathematical treatment implies a detachment from reality, that it is unable to present even an approximate picture, that it is a crude schematisation, isolating processes from their actual context. For example, two factors only are taken into account, the power of reproduction and the rapacity of species living together. None of these equations do much towards assisting the investigation of reality. If we compare Volterra's mathematical theory of the struggle for existence with the biological treatment accorded this problem by Darwin in his *Origin of Species,* the superiority of the latter admits of no doubt.

Finally we would refer to Fauré Fremiet's work: *La cinétique du développment.* He attempted to treat the laws of the growth of organic being on the basis of the differential equation and gives a brief formulation of his attitude in the following words: " L'ontogonèse peut être definie par les variations, en fonctions du temps, de deux charactéristiques, qui sont : (1) La masse du substance constituant un système organisé ; (2) le dégré de hétérogénité et la complexité physico-chimique de ce système."

The main defect in all these attempts is that they leave out of account the specific peculiarities of the given process, the given concrete, complex phenomenon (malaria, ecology, heredity), that the particular which characterises these laws and these alone, is ignored. The infinitely involved universal connection which characterises the objective reality in the given concrete form is simplified, coarsened and obliterated, is reduced to a narrow causality partially grasped, representing only a small part of the world in all its intricacy.

The diametrically opposed standpoint adopted in the critical situation prevailing in the natural sciences to-day, is expressed most clearly in that version of the so-called principle of indeterminacy given by its author Heisenberg, and his numerous adherents. This principle of indeterminacy is translated from the mathematical into human language in the following way: in principle it is impossible to determine with equal accuracy both the position and the speed of an electron. The more exact our measurement of one of these magnitudes, the less exact is the other; this is not a result of imperfections in our instruments; indeed, the contrary is true—that the process of measurement itself exercises an influence upon the position and the speed of the electron, and the more exact the measurement is, the greater the influence. Thence it follows that if we know the original state of the electron (position and speed), we cannot determine its state within any given time, from which Heisenberg draws the following conclusion (see *Zeitschrift für Physik* No. 43, 1927):

"Since all experiments are subject to the laws of quanta mechanics, the invalidity of the law of causality is definitely established by quanta-mechanics." It is clear, in this instance, that the methodological root of the error lies in the undialectical conception of the relation between the general and the particular. Heisenberg, and quanta mechanics as a whole, correctly emphasise the existence of mutual inter-action in any actual process, but cannot by themselves tackle the problem it raises. Marx, who dealt with the question of the mutual inter-action of the laws of capitalist production wrote in *Capital* (vol. III. Chap. X.):

"It is evident that the essential fundamental laws of production cannot be explained by the inter-action of supply and demand (quite aside from a deeper analysis of these two motive forces of social production which would be out of place here). For these laws cannot be observed in their pure state, until the effects of supply and demand are suspended, are balanced. As a matter of fact supply and demand never balance, or if they do, it is by mere accident, it is scientifically rated at zero, it is considered as not happening. But political economy assumes that supply and demand balance one another. Why? For no other reason, primarily, than to be able to study phenomena in their fundamental relations, in that elementary form which corresponds to their conception, that is to say, to study them unhampered by the disturbing interference of demand and supply." In eliminating the variations of demand and supply, Marx engages in a process of abstraction, but this abstraction retains the essential features of capitalism, the laws which characterise it and it alone, while an " abstraction " of the law of surplus value which retained supply and demand would divest the given form of production of its *specificum,* would resolve the particular in the general and lead to that " inadequacy and sterility of the pure concept of inter-action " of which Lenin speaks in commenting upon Hegel and demanding " intermediation (connection) in the application of the principles of causation."

The so-called mathematical pendulum is an abstraction, distinguished from a real physical pendulum in that an abstraction is made of the mass of threads and the entire mass of the swinging body is considered as concentrated upon one point. Then it is not difficult to represent the law of oscillation of such a pendulum in empty space by a typical dynamic law. Given the length of the pendulum as l, the gravitation acceleration as g and the initial amplitude as α the corresponding amplitude α can be calculated for any time t with the required accuracy. Quanta mechanics maintains, as against this, that the smaller the content in which the mass of the swinging body is concentrated, the nearer we approach to molecular, atomic, electronic dimensions, the clearer becomes the statistical character of the laws, the law of Brown's movement, that is, of the heat movement of the molecules of any given body, which runs: the average of the squares of the deflections x of each particle from its initial position is directly proportional to the product of the average of its kinetic energy $E \times$ the time t for which this average is ascertained. Consequently, conclude Heisenberg, Schlick and others, we are not in the least entitled to make the abstractions that lead to this dynamic law. Here it is quite clear that they fail to understand the general and the particular. The statistical laws of Brown's movement, to which, in the given case, quanta mechanics refers, are equally valid for the movement of any body and are therefore able to tell us as little of the movement of a pendulum as the law of supply and demand tells us of the actual essential laws of the capitalist mode of production.

In biology, as in physics, indeterminism is characterised by a failure to understand the relation between the general and the particular. The particular, the individual, the qualitative *specificum*, its uniqueness and irrepeatability is emphasised and the general is transformed into an illusion or at best portrayed as subjective. Such a methodological attitude leads to the denial of general laws, or else these laws are degraded to something subjective, relative, the product of logic. Organic processes are regarded as spontaneously originating in themselves and one of the arguments adduced in proof thereof is the periodicity of the appearance of life. Indeterminism in its most varied forms replaces causation, and we reach Verworn's conditionalism, not far removed from Machism, which replaces the explanation, the discovery of the nature of a process by a description of its conditions, its causal connections, by functional relations.

Vitalism comes into the same category, for it too denies causation in favour of teleology, or at least places limits upon it. Thus, for example, Driesch writes in the *Biologisches Zentralbratt,* 1927: "In the inorganic sphere we can make predictions if the instantaneous constellation, instantaneous velocity and fundamental law are fully known; we cannot do this in biology." Indeterminism in present day physics has exerted a powerful influence upon biology; this is shown very clearly in Bertalanffy's articles which put forward finality and teleology as against causality and the law of the conservation of energy. It is significant that both indeterminism, which denies prediction, and fatalism which in principle does not exclude it, though in fact depriving it of any significance since, by that theory, the course of events is absolutely determined beforehand and cannot be changed, are closely allied from the point of view of the theory of knowledge. Thus, for example, Eimer's orthogenesis and Berg's nomogenesis, permit predictions of further developments, which does not, however, deprive either of its vitalist character. On the other hand emergent evolution, the *Gestalt* theory etc. which have a vitalist character, maintain that evolution is quite unpredictable, indecomposable and cannot be traced back to its origin. Thus absolute necessity falls into line with absolute chance.

In biology, the greatest interest, from the methodological point of view, attaches to that variety of indeterminism, expressed in the form of statistiscal regularity, which dates back to the time of Quételet, who laid the foundations for statistics of variation. In the works of Quételet, as of Dalton, and particularly of the biometrical school of Pearson which has given mathematical form to these laws, statistical regularity means nothing but the admission that dynamic laws and, in particular, their chief factor, conditionality, do not suffice, that is, the factor of chance is recognised. It should be pointed out, however, that statistical laws, which were worked out in reaction against the mechanistic character of dynamic regularity, against the denial of quality, do themselves to a certain extent repeat the same mechanistic error: soon everything became a matter of quantity, of the average, of the purely formal enumeration of external circumstances, while the inner connections and the structure of processes were ignored and often distorted, as, for example, in Galton's famous laws.

The solution of the problem lies in the synthesis of both forms of regularity, each of which is merely a factor of the other: in reality the two exist in an internally contradictory unity, inherent in each process, each movement of matter; it is precisely their inter-penetration and struggle that embody the immanent development of matter; the mathematical expression of these laws is not adequate to reality; hence it follows that the knowledge which is based only on dynamic, or alternatively only on statistical regularity, is certain to be incomplete, one-sided, approximate and if it claims to comprehend the whole of reality, it will also be unscientific. That is why Marx, who discovered the essential laws of capitalist production, and in doing so abstracted supply and demand, declared that these laws "only operate in an extremely intricate and approximate fashion, as the average, impossible to determine accurately, of eternal fluctuations, as the prevailing tendency." (Capital. Vol. III. Chapter 9). The natural sciences to-day, convulsed by the problem of reconciling the corpuscular and the wave theory of light and matter, of reconciling continuity and discontinuity, can find a way out of this blind alley only with the aid of materialist dialectics, with its conception of every law as a unity of the inter-penetration and struggle of contradictions.

The protracted and painful labour through which the natural sciences to-day are giving birth to dialectical materialism strengthens our conviction that without the midwife of history, the revolution, the matter will not proceed, that it is only the new generation of proletarian investigators of nature, liberated from the ideological slavery of capitalism, that will finally destroy the old traditions of metaphysical methodology that now hamper science; for the essential reason that prevents the bourgeois investigator of nature, the spontaneous materialist, from recognising dialectics and which, even in the period of capitalist decline, urges him towards idealism, is his close association with the ruling class and their ideology.

Between the two extremes of bourgeois philosophy concerning the character of natural laws, between mechanism and indeterminism, there lies a vast field of dualistic and eclectic intermediate tendencies. The most typical is represented by the Machists, among them Mises, who proceed from the principle that while dynamic laws govern the individual elements forming a manifold, statistical laws govern the manifold as a whole. This they regard as the objective treatment of statistical laws, frequently expressing their outlook in the following way: statistical law is the law of the macrocosm, dynamical law the law of the microcosm. But there is no whole which cannot itself be a part, not any part that cannot be represented as a whole manifold; our earth is a macrocosm in comparison with the molecules of which it is composed, but a microcosm, in comparison with the stellar system of the Milky

Way. Thus the objective in the definition of the character of regularity is lost, and everything depends upon how we regard the given object. On the other hand, this definition is metaphysical, for dynamic and statistical laws are represented as existing apart from each other. Hence it follows that the eclectics' struggle against the mechanistic conception of statistical regularity as a provisional substitute for our ignorance (since finally the law of the whole is recognised to be merely the sum of individual laws), is conducted from a formalist standpoint. While conceding to the mechanists that the relation between statistical and dynamical regularity is the same as the relation between the law of the whole and the law of the parts, they exclaim against the attempts of the mechanists to reduce the laws of the whole to the laws of the parts. Eclectics draw a distinction between additive and non-additive properties, and maintain that it is the non-additive properties, which are not peculiar to each separate element of the manifold, but attach to it as such, that are portrayed by statistical laws; for this reason the statistical laws of the whole cannot be ascertained by the quantitive summing up of the laws of single events. This argument, however, does not achieve its purpose. Mechanists may dissociate themselves completely from an identification of the laws of the whole with the simple sum of the laws of the individual parts—they are still at their former position, in which they maintain that quality, can be reduced to quantity.

Actually, what meaning is there in the assertion that, for example, the statistical laws of the state of gas as a whole, cannot be reduced to the multiplicity of the dynamic laws of its molecules? This is a case of a non-additive manifold: all molecules are in process of inter-action, such properties of the whole as temperature, pressure and volume cannot be developed as the simple sum of the properties of separate molecules. But does that mean that it is altogether impossible to develop the laws of the whole on the basis of the dynamical laws of movement of the individual parts? The question itself engenders agnosticism. In fact, the process of inter-action exists; connections between the particles are formed; why then should our knowledge fail to follow the process of origin? Perhaps because we cannot tackle the matter mathematically? That is the answer given by the mechanists: if we could solve the differential equations of the movement of the individual parts, we should thereby comprehend the movement of the whole and consequently, in principle, the laws of the whole can be reduced to the laws of the parts. Not at all! We can adduce examples in which, in the case of the simplest mechanical movement, we are able to resolve the system of the differential equations of a non-additive manifold. But what does that give us? The knowledge of the behaviour of each individual particle, but that still does not tell us anything of the behaviour of the mechanical manifold as a whole.

Further, if we could indicate a way of ascertaining, from a knowledge of the laws of all the parts, the laws of the whole, we should still not be able to speak of reduction, for, first of all, this would certainly introduce a new quality, and secondly it would be inconvertible: from the laws of the parts we would have ascertained full quantitative knowledge of the laws of the whole, but the contrary does not hold. Such a way is indicated by passing to the limit. The distribution of the characteristic magnitudes (position, speed, etc.), approaches, with exceeding over all limits, that distribution which can be ascertained as a statistical regularity by the calculation of probability. It is also to be expected, therefore, that dynamic laws are the laws of the particular, the laws of the quality which gives its general character to statistical regularity.

" In *Capital*, Marx first analyses the simplest, most commonplace, fundamental relation of bourgeois economy, encountered a million times—the exchange of commodities. In this simple phenomenon (in this " cell " of bourgeois society) the analysis covers all contradictions (or the germs of all contradictions) in present-day society. Further on we are given the development (and growth and movement) of these contradictions and this society, of their individual parts, from beginning to end." In this way, according to Lenin, the dialectical method does not rest solely upon dynamic regularity (as the mechanists would have it), nor solely upon statistical regularity (as suggested by those who would solve the problem in a formalist-idealist fashion). The dialectical materialist will conduct his investigations, not with the object of replacing statistical by dynamical laws, but in order to comprehend the object in the internally-contradictory unity of its content and its form, of the particular and the general, of the accidental and the necessary, of the discrete and the continuous. Our course lies not in " maintaining the old conception of necessity " and " forcing upon nature, in the form of a law, a logical construction contradicting both itself and reality," nor in " declaring the chaotic realm of accident to be the sole law of living nature," (Engels, *Naturdialektik*) nor in making an eclectic choice, but " in showing that the Darwinian theory represents in fact the correctness of the Hegelian conception of the essential connection between necessity and chance," in showing that " the division of the single and the understanding of its opposing parts is the essence of dialectics," " in examining the correctness of this aspect of the content of dialectics by a study of the history of science," in which we conceive of the " identity of contradictions " not only as " a sum of examples " but as "the cognitive law (and the law of the objective world)."

Both in theory and practice the question of the limits of application of laws is of particular importance. Lenin's contention that " the conception of law is merely a stage in the human understanding of the unity and connection, the mutual dependence and uniformity of the world process," his statements that, particularly in present day physics, " a struggle must be conducted against making the idea of law absolute and primitive, against giving it the character of a fetish," that, in phenomena, the law selects the calm and consequently, every law is partial, incomplete, approximate, all this refers both to dynamic and to statistical laws, for every example of the general is from another aspect an example of the individual. Everything depends upon the understanding how to determine in the given case and in the given circumstances, the limits of the general, upon avoiding a complete break with what forms the essence of the phenomenon, and falling into empty abstraction. Such moderation can be acquired only by studying concrete conditions and their transformations, only by practice, of which Lenin said:—" Practice stands higher than (theoretical) knowledge, for it possesses the distinction not only of general validity, but also of direct reality." In statistics, which is concerned with the laws of the general, the question of limits is often forgotten. In the writings of his youth : *Who are the friends of the People; The development of Capitalism in Russia; A new Economic Movement in Peasant Life;* Lenin pointed out to bourgeois statisticians, frequently and in detail, how little scientific value attaches to the general average, the fictitious average, how statistics becomes a game with figures when, for example, the farms of poor peasants are added to the farms possessed by peasants who employ wage labour and to the farms of the large landowners and the total is divided by the total number of farms, etc., etc. Stalin, in a speech delivered in April, 1929, developed this idea by applying it to conditions in the Soviet Union at that time, conditions which have already been completely changed. Dealing with the extent of the area sown, he said:—" The method of averages if not corrected by the data of the different districts, is not a scientific method. In the *Development of Capitalism* Lenin criticises the bourgeois economists who apply the method of averages without referring to the facts of the different districts under cultivation. If we consider the movement of the area sown, that is, if we consider the matter scientifically, we see that in some districts this area is steadily increasing, in others, often because of meteorological conditions, falling, although nothing points to a steady decrease in the area sown anywhere." Statistical laws lose their scientific value, if the essential aspect of the phenomenon is forgotten, if the particular is metaphysically denied instead of being dialectically handled.

The most important factor in the unity of dynamic and statistical regularity is the direction of the movement of the law which, whether manifested as a dynamic or a statistical law, is in process of self-movement. In which direction do the internally contradictory tendencies develop? To follow the direction of the development of a law we have to consider not the process as a whole, for there the direction is expressed always as a negation of negation, but, as Engels pointed out in Anti-Dühring, we must study concretely this negation of negation.

Only by investigation in the different spheres of science conducted by dialectical materialists from this standpoint, shall we be able to work out the problem of the conditions in which the opposing factors of a law melt into each other without running the danger " of playing a futile game with empty analogies, of falling into abstruse Hegelianism," against which Lenin uttered a warning. Each of the brilliant concrete examples of the negation of negation adduced by Marx, Engels and Lenin will form the backbone for researches into all the new material offered by the most recent developments in society and its ideology, including both physics and biology.

THE PROBLEM OF THE ORIGIN OF THE WORLD'S AGRICULTURE IN THE LIGHT OF THE LATEST INVESTIGATIONS.

By N. I. VAVILOV.

THE PROBLEM OF THE ORIGIN OF THE WORLD'S AGRICULTURE IN THE LIGHT OF THE LATEST INVESTIGATIONS.

By Prof. N. I. VAVILOV.
Member of Academy of Sciences of USSR.
President of the Lenin Academy of Agricultural Sciences.

Where are the beginnings of agriculture to be sought? Were they independent in different regions, in different continents? How is the geographical localisation of primitive agriculture to be explained? Which plants were first brought into cultivation? Which animals were first domesticated, and where? Where shall we find the primary sources of cultivated plants? How are modern domesticated field animals and cultivated plants connected with their wild related types? How did the evolution of cultivated plants and animals proceed? How are primary agricultural civilizations connected? Which implements were used by primitive agriculturists in different regions?

Viewed from the standpoint of concrete materialistic studies all these historical questions are very actual, and of great significance for modern agriculture. In contradistinction to past practice, the present-day investigator, faced with increasingly difficult economic conditions in the world, attempts to utilise the experience of the past in order to improve upon existing practice. In the Soviet Union, which is now building up socialism and socialistic agriculture, we are interested in the problem of the origin of agriculture, and of the origin of cultivated plants and animals chiefly from the *dynamic* viewpoint. By knowledge of the past, by studying the elements from which agriculture has developed, by collecting cultivated plants in the ancient centres of agriculture, we seek to master the historical process. We wish to know how to modify cultivated plants and domestic animals according to the requirements of the day. We are but slightly interested in the wheat and barley found in the graves of Pharaohs of the earliest dynasties. To us, constructive questions—problems which interest the engineer—are more urgent. It is much more important for us to know how Egyptian wheat differs from wheats of other countries, which characteristics in this Egyptian wheat are of importance in order to improve our wheat, to understand how this Egyptian wheat has originated. The investigator wishes to find the primary elements, " the bricks and mortar," from which the modern species and varieties were created. We need this knowledge in order to possess the initial material for practical plant and animal breeding. We study the construction of primitive agricultural implements in order to get indications for the construction of modern machinery.

In brief, the historical problems of this origin of agriculture, of the origin of cultivated plants and domesticated animals are especially interesting for us in the sense of mastering and controlling the breeding of cultivated plants and animals.

The results of these studies may be of interest to archæologists, historians, naturalists, agronomists, geneticists, plant and animal breeders. Therefore we take the opportunity to-day at this International Congress devoted to the history of science and technology, to draw your attention to the chief results of investigations into this subject, which have been made recently in the Soviet Union.

In the course of our work on the practical questions connected with plant-breeding, we have approached some of the problems of the world history of agriculture included in this Congress programme.

The Institute of Plant Industry in Leningrad has recently been studying the cultivated plants of the whole world according to a definite programme. During the systematic study of a number of species it became evident that so far neither the botanist, nor the agronomist, nor the breeder has yet, with any degree of completeness, approached the study of the world's resources even of the most important cultivated plants, whose centres of evolution, as investigations have shown, are located chiefly in ancient agricultural countries. Contemporary European and American horticulture and agriculture know only fragmentary details, derived from ancient centres of agriculture, of the initial diversity of cultivated plants.

We began to study systematically the cultivated plants of the world. Numerous special expeditions were sent to different parts of the globe, chiefly to ancient mountainous countries. They collected an enormous amount of material and new data about the primitive ways and technique of agriculture. The investigations embraced the countries of the Mediterranean, including Morocco, Algeria, Tunisia, Egypt, Portugal, Spain, Italy, Greece, the whole of Asia Minor, Syria, Palestine and the islands of Sicily, Sardinia, Crete, Cyprus and Rhodes. In detail were investigated: Abyssinia, Eritrea, Persia, Afghanistan, Western China, agricultural Mongolia, Japan, Korea, Formosa, and to some extent India. The ancient agricultural regions of Transcaucasia and Turkestan were studied most closely. In the new world the investigations embraced the whole of Mexico (including Yucatan), Guatemala, Columbia, Peru, Bolivia and Chili.

These expeditions collected a great number of specimens of cultivated plants (hundreds of thousands), which have now been studied for several years at different experimental stations. The investigations elucidated the world's geographical distribution of species and varieties; they discovered many specimens so far unknown to botanists, breeders and agronomists, often having valuable "practical qualities." They led even to the discovery of *new species* of cultivated plants. Thus In Peru and Bolivia our expeditions discovered twelve new species of potatoes, instead of the one known species (*Solanum tuberosum*). New species of wheat and thousands of new varieties of small grains and of other field and vegetable plants.

The most essential fact established by these investigations, one which is of great importance to the comprehension of the history of the world's agriculture, is the geographical localisation of the chief varieties of cultivated plants. This has been established by close observation. It has been proved possible to locate exactly the primary original centres of the most important cultivated plants, for instance, of wheat, barley, rice, maize, of many field and vegetable crops. This facilitated the acquiring of an enormous amount of basic material, hitherto unknown to botanists.

The fundamental centres of origin of cultivated plants, as was proved by these investigations, very frequently play the rôle of accumulators of an astonishing diversity of varieties. In small, primitive, agricultural Abyssinia alone, where the whole area under wheat is certainly no more than half a million hectares, we found more varieties than in all the other countries of the world taken together. The varieties of maize in Southern Mexico—the initial home of this plant—are extremely rich. The wild fruits in Transcaucasia—the chief home of many European fruit trees—are astonishingly varied. Diversity of varieties, however, alone, does not always determine the primary centre of origin of the cultivated plant. It is necessary to study their wild and cultivated stocks, the history of the plant's migrations. We have elaborated methods of differential systemization and of botanical geography which allow us to determine exactly the initial home of single cultivated plants.*

As a result of the investigation of several hundreds of cultivated plants we succeeded in establishing the fundamental world centres of the chief cultivated plants. Some of these results are probably of general interest.

In general our investigations have led to the establishment on the earth of seven fundamental, independent centres of origin

*N. I. Vavilov.—Studies on the Origin of Cultivated Plants. Bull. of Applied Botany, Vol. XVI No. 2. 1926.

N. I. Vavilov.—Regularities in the Geographical Distribution of Genes of Cultivated Plants. Ibid. Vol. XVII. 1927.

N. I. Vavilov.—Mexico and Central America, as a fundamental centre of the origin of cultivated plants of the New World. Ibid. 1931.

N. I. Vavilov.—The Linnean Species as a system. Ibid. 1931.

of cultivated plants, which at the same time were the probable foci of the independent development of world agriculture.

For the majority of our present cultivated plants the chief continent is Asia. A great number of cultivated plants are of Asiatic origin. In Asia we distinguish three fundamental centres of species formation. First and foremost, South-Western Asia, including the interior of Asia Minor, Persia, Afghanistan, Turkestan and North-Western India. Here is the home of soft wheats, of rye, flax, alphalpha, Persian clover (*Trifolium resupinatum*), of many European fruit trees (apple, pear, *Prunus divaricata,* pomegranate, quince, sweet cherry), of grapes, of many vegetables.

It is not altogether by chance that Biblical history locates the primary paradise, the Garden of Eden, in this region. Even now it is possible to see forests of wild apples, pears, sweet cherries, quinces, covered with wild grape-vines—paradises in the full sense of the word, in Transcaucasia and in Northern Persia.

The second independent world centre in Asia is located in India proper, including the valley of the Ganges, the whole of Indostan peninsula, and the adjoining parts of Indochina and Siam. This is the original home of rice—the most important crop in the world—which is still the staple food of half of all mankind. Here it is still possible to observe rice in its primary stage as a wild plant, as a weed in the fields, and to follow its development into the primitive cultivated forms, which display an astonishing diversity. Here also is the home of many tropical cultivated plants, sugar cane, Asiatic cottons, tropical fruit trees (for instance, the mangoes).

The third Asiatic centre is located in Eastern and Central mountainous China. Central Asia, as we now know, beyond question, had no relation to primary agriculture, notwithstanding its immense territory. Neither Mongolia nor Western China, Tian-shan, nor Siberia shows any traces of independent agriculture, whether in regard to the diversity of crop plants, or to the technique of agriculture.*

Eastern Asia, on the contrary, the upper course and the valleys of the great rivers of China, Hun-ho and Yangtze-Kiang, have given birth to the great Chinese culture, and perhaps even to pre-Chinese agriculture. This is the home of many plants, such as the peculiar Chinese cabbages, the radish, and of many peculiar Chinese crops little known in Europe. This is the native country of Citrus plants, Unabi (*Zizyphus*), the persimmon, the peach, the Chinese plum (*Prunus Simoni*), the tea-shrub, the mulberry tree, of many tropical and especially sub-tropical plants.

*N. I. Vavilov.—The rôle of Central Asia in the origin of cultivated plants. Bull. of Appl. Botany, 1931.

In this country the technique of agriculture is very peculiar. The soil is cultivated chiefly by hand labour, farm animals rarely being used. The intensive cultivation of truck crops is widely spread. Rain in China is brought by the monsoons. The chief agricultural regions are supplied with an adequate amount of moisture. As to Japan and Formosa, our investigations have shown that these countries have borrowed their crops and technique from China. The same may be said of the Philippines and of the Malayan islands, whose agriculture is chiefly borrowed from China.

In contradistinction to China and Japan, South Western Asia (the first centre) is characterized by an extensive use of farm animals—cattle, horses, camels, and mules. The diversity of agricultural implements is here especially noteworthy.

In Europe, primary agriculture is definitely confined to the South. The *fourth world centre* embraces the ancient countries adjoining the Mediterranean, including the Pyrennean, Appenine and Balkan peninsulas, the coastal region of Asia Minor, Egypt, and also the territory of modern Morocco, Algeria, Tunisia, Syria and Palestine.

In spite of the great historical and cultural importance of the Mediterranean centre, which has given rise to the greatest civilizations of antiquity—the Egyptian, Etruscan, Aegean and the Ancient Hebrew—this centre, according to the investigation of its varietal diversity, includes but few autochtonically important crops. Ancient agriculture is here based on the olive, the carob-tree (*Ceratonia siliqua*), the fig tree. The majority of field crops, such as wheat, barley, beans and peas, have obviously been borrowed from other centres. The varietal diversity of the crops is here considerably poorer than in the principal centres of the corresponding crops. Only a series of forage plants, such as *Hedysarum coronarium,* the forage lens, *Ervum Ervilia,* the forage vetches, *Lathyrus cicer, L. Gorgonia, Trifolium alexandrinum,* have originated in the Mediterranean region.

Here the cultivated plants have undergone a careful selection promoted by the mild climate and the high level of culture of the population. The varieties of cereals, leguminous grain crops, flax, truck plants, are distinguished in the Mediterranean region by an extraordinary large size of fruits, seeds, bulbs, as well as by their fine quality, by comparison with the corresponding crops growing in regions distant from the Mediterranean.

The primeval agriculture of the Mediterranean is characterized by special types of implements for tilling as well as for harvesting such as the Roman furrow plough, the threshing board set with sharp stones, and the stone roller. China, India, and to a considerable extent, South-Western Asia, are not acquainted with these types of implements.

The fifth world centre is found in mountainous Eastern Africa, chiefly in mountainous Abyssinia. This small centre is rather peculiar, being characterized by a small number of independent important cultivated plants displaying an extraordinary variety of forms. Here we find the maximum diversity in the world, so far as the varieties of wheat, barley, and perhaps also the grain sorghum, are concerned. This is the home of such plants as teff-*Eragrostis abyssinica*: a most important cereal of Abyssinia; *Noug-Guizotia abyssinica*: an original oleaginous plant of this country. Flax is distinguished by its small seeds. In distinction from the ancient Mediterranean countries and South-Western Asia, it is grown in Abyssinia only as a bread plant for the sake of its flour. The cultivation of flax for oil and fibre is still unknown to primitive Ethiopia. Abyssinia is the home of the coffee plant as well as of the barley used in brewing.

Though no archaeological memorials testifying to the ancient character of the Abyssinian centre have been found (with the exception of the ancient phallic culture recently discovered in Southern Abyssinia), it may be affirmed, on the basis of the diversity and the special qualities of the cultivated plants, as well as the technique of agriculture (cultivation by means of hoes still exists to some extent in Abyssinia), that this centre is indubitably independent and very ancient. It is our conviction that Egypt has borrowed its crop plants from Abyssinia to a considerable extent. All comparative data concerning the diversity of the cultivated plants, the species of domestic animals, the mode of life of the agricultural population, the original food of the latter—all point to the autonomous character of the Abyssinian centre. The data of linguistics are a further proof of the foregoing conclusion.

In the New World investigated during the last five years by Soviet expeditions, two principal centres must be distinguished: the South Mexican centre including part of Central America, and the Peruvian including Bolivia. The former is the more important. It has given rise to such crops as corn, the Upland cotton, cacao, agave-hennequen, the musky pumpkin, the multiflorous and the common bean, chiota (sechium) papaya, and many indigenous crops of secondary importance*.

Peru and Bolivia are the home of the potato, the chinchona tree, the coca shrub, as well as of a series of secondary crops. Here extraordinarily polymorphic groups of the soft corn have been differentiated.

The other regions of South and Central America, though they have given rise to some crops, are of no decisive importance for the history of world agriculture*.

*N. I. Vavilov.—Mexico and Central America, as the principal centre of origin of the cultivated plants of the New World. Bull. of Appl. Botany, 1931.

The agricultural centres of the New World have come into existence quite independently of those of the Old World, a fact proved by the unique cultivated flora of North and South America. The ancient civilizations of the Mayas and the Incas did not know the use of iron, were not acquainted with the plough. The " foot plough " known in the high mountain regions of Peru, is, after all, no more than a spade. Neither Mexico nor Peru had farm animals for agricultural purposes. The llama and the alpaca, as well as the guinea-pig, domesticated in Peru, were raised for the sake of meat and wool, and only the former served as a beast of burden.

Such are the seven principal centres of the world, that have given rise to the whole world agriculture. As may be seen from the appended map, these centres occupy a very limited territory. According to our estimates the Mexican centre in North America occupies about 1/40th of the whole territory of the vast continent. About the same area is occupied by the Peruvian centre in regard to the whole of South America.

The same may be said about the majority of centres in the Old World. The differentiation in the types of agricultural implements corresponds to the differentiation in the primary centres of origin of the cultivated plants. In mountainous East Africa, as well as in the whole of primitive Africa, the cultivation of soil with a hoe may be observed even to-day. As the investigation of plough agriculture throughout the world, made by B. N. Zhavaronkov has shown, the ploughs of Abyssinia, China, South-Western Asia, and of the Mediterranean countries, are of different types.

The geographical location of the primary agricultural centres is rather peculiar. All seven centres are chiefly confined to the tropical and sub-tropical mountain regions. The centres of the New World are confined to the tropical Andes, those of the Old World to the Himalaya, the Hindu-Kush, mountainous Africa, the mountain regions of the Mediterranean countries, and to mountainous China.

After all, only a narrow strip of the dry land of the earth has played an important rôle in the history of world agriculture.

From the point of view of dialectics, considered in the light of the latest investigations, the geographical concentration of the great primeval agricultures in this limited zone, becomes comprehensible. The tropics and the sub-tropics provide optimum conditions for the unfolding of the process of species origination. The maximum species diversity shown by the wild vegetation obviously gravitates towards the tropics. This is especially conspicuous in North America, where South Mexico and Central America, occupying a relatively inconsiderable area, contain more plant species than the whole vast expanse of Canada, Alaska, and the United States (California included) taken together. The republics of Costa Rica and Salvador, pigmies in the sense of the territory they occupy, nevertheless supply the same number of species as the United States, 100 times their size. The powerful process of species origination is geographically localized chiefly towards the moist tropics of the New World.

The same may be observed in the Old World. The Mediterranean countries are very rich in species. The flora of the Balkan peninsula, Asia Minor, Persia, Syria, Palestine, Algeria, Morocco, is distinguished by a great multiplicity of species (4,000-6,000 species on an average), exceeding in this regard Northern and Central Europe. India possesses not less than 14,500 species. The flora of Central and Eastern China displays an extraordinary diversity. Though the more or less exact number of species shown by this most interesting part of China is not known, it runs into many thousands.

Abyssinia is rich in indigenous plants, as well as by its number of species in general.

Thus, the geographical location of the species and of the form origination of cultivated plants coincides, to a considerable extent, with the location of the general process of species origination shown by the floras of the world.

The processes of mountain formation indubitably have played an important rôle in the differentiation of the vegetation into species, promoting the process of the divarigation of species. Isolators, mountain barriers checking the spread of species and genera have been of great importance to the differentiation of separate forms and whole species. The various climates and soils found in the mountain zones to which gravitate the principal centres of the origin of cultivated plants, promote the development of diversity among the species, as well as within the species composition of these plants. On the other hand, the glaciers which in the preceding geological epoch covered Northern Europe, North America and Siberia, have destroyed whole floras.

If the moist subtropics chiefly favour the development of trees, the tropical and subtropical mountain regions, where the primeval agricultures settled, are characterized by the development of herbaceous species, to which belongs the majority of the most important plants of the earth.

The tropical and subtropical mountain regions afford optimum conditions for human settlement. Primeval man was afraid, and is still afraid, of the moist tropics, with their luxuriant vegetation and their tropical diseases, though the vast subtropics with their highly fertile soils occupy one-third of the dry land on the whole earth (Sapper). For his domicile man turned, and continues to turn, to the borders of the tropical forests. The tropical and subtropical mountain regions offered the most favourable conditions of warmth and abundance of food to the first settlers. In Central America and Mexico man still utilizes a multitude of wild plants. It is not always easy to distinguish the cultivated from their corresponding wild plants.

The mountain contour favoured life in small groups; it is with this phase that the development of human society begins. There is no doubt that the conquest of the vast basins of the Lower and Middle Nile, of the Euphrates, Tigris and Indus could be accomplished only by a population united into large groups, and this could have taken place only in the later stages of the development of human society.

The primitive man, the primitive farmer, used to live, and still continues to live in inconsiderable, isolated groups, and for him the mountainous tropics and subtropics presented exceptionally favourable conditions.

In opposition to the common views of the archaeologists, our investigations of the ancient agricultures have led us to the conclusion that primitive agriculture was not irrigated. The analysis of the diversity shown by the cultivated plants in Egypt, Mesopotamia, the irrigated regions of Peru (up to 11,000ft. above sea level) have shown that the cultivated plants of these countries have been borrowed from elsewhere. The indisputably most ancient crop plants of Abyssinia, mountainous Mexico and Central America, high mountainous Peru (above 11,000ft.), China, India, and the Mediterranean countries, were *not irrigated*.

Taking into consideration the interaction of opposed factors, and basing our deductions on concrete facts which may be verified by a direct study, we have been able to fix the exact geographical location of primitive agriculture and have determined the essential features of this localization.

It is obvious that these cultures, based on different genera and species of plants have come into existence autonomously, whether simultaneously or at different times, and that one must speak at least of *seven* principal cultures or, more exactly, groups of cultures. To them correspond quite different ethnological and linguistic groups of peoples. They are characterized by different types of agricultural implements and domestic animals.

This knowledge of the initial centres of agriculture, throws light on the whole history of mankind, and the history of general culture.

Our investigations have shown that during the spread of the cultivated plants towards the North and into the high mountain regions, the principal crops were sometimes supplanted by their attendant weeds, when the latter carried a certain value for the farmer.

Thus, the winter wheat, when migrating to the North from its chief centre of origin, South Western Asia, was supplanted in a series of regions of Asia and Europe, by the hardier weed, winter rye. In the same way barley and emmer were supplanted by oats—weeds, less exacting in regard to soil and climate. Flax is not infrequently supplanted in Europe by the weed *Camelina*, in Asia by *Eruca sativa*, and so on.

Thus, a series of crop plants has originated independently of the will of man, owing to natural selection. In studying the rye weed—mingled with wheat in South Western Asia—we have discovered a striking diversity of forms, of which the European rye-growing farmer has no idea.

A series of regularities in the succession of crops during their northward spread has been established.

We give here only a summary of our collective investigations. They have led us to the mastery of the world resources in species, and to a comprehension of the evolution of the cultivated plants, as well as to the solution of questions concerning the autonomy of the principal agricultures and their interrelations. It is natural that the centres of the New World should be more closely connected with one another than with those of Eurasia. The South Western Asiatic centre is especially near to that of Abyssinia. One has given rise to the soft wheats, the other to the hard varieties.

These data are the material prerequisites to a comprehension of the first phases of the evolution of human society. It is natural that one of the principal factors in the first settling of mankind should have been the distribution of the natural food resources.

The data appertaining to the primary geography of cultivated plants and of their wild growing relatives fit in with our present knowledge of the evolution of primeval man. South Western Asia and mountainous East Africa were evidently the original area for the creation of a human society engaged in agriculture. Here we observe the concentration of the chief elements necessary for the development of agriculture.

Such is the problem of the origin of agriculture, viewed in the light of modern methods of investigation. In approaching this problem from the point of view of dialectic materialism, we shall be led to revise many of our old concepts and, which is fundamentally important, we shall gain the possibility of controlling the historical process, in the sense of directing the evolution of cultivated plants and domestic animals according to our will.

THE WORK OF FARADAY AND MODERN DEVELOPMENTS IN THE APPLICATION OF ELECTRICAL ENERGY.

By W. TH. MITKEWICH.

THE WORK OF FARADAY AND MODERN DEVELOPMENTS IN THE APPLICATION OF ELECTRICAL ENERGY.

By W. TH. MITKEWICH.

1. The history of science frequently exhibits a close connection between the achievements of scientific research and their practical applications. In other words, we can clearly state a relation between science and technology, and acknowledge their inner unity. In this respect, we have an extremely striking example in the conditions arising from and the development of modern electro-technique. The scientific discoveries of Faraday were a mighty stimulus for the use of electrical energy in practical life.

A century ago, Faraday discovered the electro-magnetic induction of current. Thanks to his intuitive capacity, that has never been surpassed, he could look into the very nature of things, and arrive at once at a correct and clear understanding of all that was going on in the real world of electro-magnetic phenomena which surrounds us. It enabled mankind to adopt the most convenient and simple method of transforming mechanical work into electrical energy capable of easy and speedy transportation to great distances, and to many other useful transformations. Thus we can state that Faraday's discoveries were the basis of up-to-date electrical engineering, and all the applications of electrical energy. It is clear, of course, that the growth of new ideas arising from Faraday's scientific work, the construction of electro-magnetic machinery and other apparatus embodying these ideas in practical use, brought to life in this great domain the work of a number of other physicists and a number of inventors. But it is undeniable that Faraday was the true founder of this branch of applied science. The whole army of those who worked theoretically and practically, and carried on Faraday's task, was always and invariably inspired and supported by the genius of that great man.

2. The fundamental thought that guided Faraday's investigations, and led him to the discovery of electro-magnetic induction, was that between the phenomena of electricity and those of magnetism there must exist some close connection. He possessed an intuitive bent of mind that enabled him to inquire into the relationship of phenomena. Convinced of the correlation of forces and of the conservation of energy long before either of those doctrines had received distinct enunciation as principles of natural philosophy, he seems never to have viewed an action without thinking of the necessary and appropriate reaction. He seems never to have deemed any physical relation

complete in which discovery had not been made of the converse relation for which instinctively he sought as in the case of a copper coil and an iron core. Given that electricity was flowing through the one, it set up magnetism in the other. What was the converse? Searching from all angles for a solution of this question and continually varying his experiments Faraday was making his way to his aim to " Convert magnetism into electricity." At last, in the autumn of 1831, he solved the problem. He succeeded in generating the electric current by means of electro-magnetic induction! He saw from the beginning that peculiar properties of magnetic flux are manifested in this phenomenon. It is necessary to state that the conception of the magnetic flux, as such, belongs fully to Faraday himself. Truly, we must acknowledge that it was Faraday who was the founder of the doctrine of physical properties of magnetic flux. There were other physicists who saw centres of forces acting at a distance. Faraday in his mind's eye saw physical lines of force traversing all space. From mathematical fictions which were used and still continue to be used by some, Faraday's point of view leads us to a closer contact with what is actually going on. The experimental methods given by Faraday permit us, in the real sense of the word, to feel the invisible magnetic flux as something real. Faraday was the first who touched the really existing magnetic flux which has primary importance in all the manifestations of the electric current; magnetic flux bearing all the electric current energy; magnetic flux that has such an important role in all applications of the electrical energy. Faraday was the first who realised the insufficiency, the complete onesidedness, and even the fallibility of our usual conceptions on the electric current, those conceptions being connected, owing to purely historical conditions, with the process of movement of electrical fluids. He directed scientific research to the space round the conductor carrying the current, to that space where is located the electric current energy.

Faraday's thought has quite an exclusive penetration into the depth of things when analysing the specific importance of the magnetic flux in all electro-magnetic phenomena, and particularly in the electro-magnetic complex called by us the electric current flowing through some conductor. Faraday's mind was possessed more and more by this idea, which completely dominated his mind to the end of his scientific activity. All the works of Faraday's final period, beginning with series XIX of his " Experimental Researches in Electricity " are, in fact, nearly all consecrated to these questions. Clerk Maxwell, the great interpreter of Faraday's ideas, used a great deal of material from this analysis in electro-magnetic phenomena. But it is undoubtedly true that many of Faraday's scientific achievements are not yet sufficiently understood and appreciated. His " Experimental Researches in Electricity " remains an Arabian book under seven seals for those who, owing to the excesses of purely formal methods of investigation, have lost, in some degree, the capacity of understanding thoughts expressed in simple words. Faraday gave us the highest model of what physical thinking should be. He was a true natural philosopher. Every deviation from Faraday's method of study and analysis of physical phenomena leads to painful results. The roots of the modern crisis in physics must be sought to a great degree in this direction.

3. Faraday's nonformal treatment of physical phenomena, that we have just spoken of, was the intrinsic cause of the exclusive practical fruitfulness of his discoveries in electro-magnetic induction. Faraday himself settled the fundamental laws of electro-magnetic induction. At the same time physical conceptions, in a high degree adequate to their real nature, conceptions that he formulated as the basis of understanding of this phenomenon—opened an easy way for practical applications of his discovery.

The conditions for Faraday's experiments were very simple, and, generally speaking, they did not exceed purely laboratory investigation limits. And yet we meet here with all the principal elements of to-day's electrical installations. Perhaps this statement will seem somewhat exaggerated, and not quite true. But, in fact, it is quite right, and becomes entirely clear when we analyse unprejudicially the work of Faraday.

Faraday has stated that always when the conductor is moving across magnetic lines a tendency (electro-motive force) develops in this conductor, and electric current is caused if the conductor forms a part of some closed circuit. He realised during his studies several arrangements in which he generated by these means an electric current of alternating directions. In other arrangements by applying a copper disc rotating between magnet poles he obtained a constant electric current. All this we can and must consider as a prototype of modern dynamo-electric machinery. Even such an important part as the dynamo commutator may be seen in a rudimentary form in his experiment of the rotating disc with a sliding brush on its edge, this brush taking off the current from its radial elements passing under the brush in succession.

In Faraday's experiments with an iron ring, having two separate windings, one of which (primary) was alternatively joined to or cut off from the battery and the other (secondary) was connected with a galvanometer by means of sufficiently long conductors, he gave us the prototype of a modern alternating current transformer—a most important part in every power transmission and distribution system.

The conductors connecting the electric generator to the galvanometer or to other Faraday's receivers of electrical energy are the prototype of modern electric power transmission line.

The galvanometer used by Faraday, or the minute spark between the lightly touching charcoal points were the prototype of modern electrical energy receivers. We can find in a galvanometer the simplest electromagnetic mechanism transforming electrical energy into motion. It performs the same function as does any modern electromotor uninterruptedly and in a more perfect way. The prototype of such an electro-motor with a constant rotary motion we can see in Faraday's experiments, in which he discovered that a wire included in the circuit, but mounted so as to hang with its lower end in a pool of quicksilver, could rotate around the pole of a magnet; and conversely that if the wire were fixed and the pole of the magnet free to move, the latter would rotate around the former.

The first steps in the practical application of electric lighting ought to be connected with the name of Faraday. He was a permanent consultor for many long years at Trinity House on different questions and, in particular, concerning the feeding of electric arc lamps from magneto-electric machines. To Faraday belongs the idea of using for electric lighting an incandescent lamp containing a platinum wire spiral.

Thus we see how gigantic was the work of Faraday in the sphere of electro-magnetic phenomena and how much he contributed to the modern development of the applications of electrical energy.

4. In conclusion it is very interesting to read the following passage from Clerk Maxwell's article on Faraday in the "Encyclopædia Britannica," which admirably sums up the matter :—

"The magnitude and originality of Faraday's achievement may be estimated by tracing the subsequent history of his discovery. As might be expected it was at once made the subject of investigation by a whole scientific world, but some of the most experienced physicists were unable to avoid mistakes in stating, in what they conceived to be more scientific language than Faraday's, the phenomena before them. Up to the present time the mathematicians who have rejected Faraday's method of stating his law as unworthy of the precision of their science, have never succeeded in devising any essentially different formula which shall fully express the phenomena without introducing hypotheses about the mutual action of things which have no physical existence, such as elements of currents which flow out of nothing, then along a wire, and finally sink into nothing again.

"After nearly half a century of labour of this kind we may say that, though the practical applications of Faraday's discovery have increased and are increasing in number and value every year, no exception to the statement of these laws as given by Faraday has been discovered, no new law has been added to them, and Faraday's original statement remains to this day the only one which asserts no more than can be verified by experiment, and the only one by which the theory of the phenomena can be expressed in a manner which is exactly and numerically accurate, and at the same time within the range of elementary methods of exposition."

ELECTRIFICATION AS THE BASIS OF TECHNICAL RECONSTRUCTION IN THE SOVIET UNION.

By M. RUBINSTEIN.

ELECTRIFICATION AS THE BASIS OF TECHNICAL RECONSTRUCTION IN THE SOVIET UNION.

By Prof. M. RUBINSTEIN.

The Soviet Union has set itself the task of accomplishing the Five-Year Plan in four years, and in a number of essential, decisive branches of economy, even in three years.

By 1932 the Five-Year Plan must be fully carried out, and, in a number of branches, with a substantial margin.

This confronts us with the task of starting to work right away on the drafting of a new perspective for the 2nd Five-Year Plan. In this plan, which maps out the course for the development of socialist relations in production based upon the completion of the collectivization of agriculture, and upon the necessity of fully solving the problem of "technically and economically overtaking and outstripping the advanced capitalist countries," an important place must be occupied by the plan of technical reconstruction, by the solution of the problem of "covering in no more than ten years the distance by which we lag behind the advanced countries of capitalism." (Stalin.)

Upon the basis of the unprecedented experience of the past ten years, this problem is now being taken up by the Soviet Union on an infinitely larger scale, and in far more real and concrete fashion, than could be done at the commencement of the reconstruction period of our economy.

As regards technical reconstruction, the new plan is *fundamentally* based upon the modern achievements of capitalistic technique throughout the world (and, in the first place, in America), while taking stock of the clearly discernible ways of its further development, and in combination and conjunction with the full development of socialist conditions of production.

As was repeatedly emphasized by Lenin, we must "combine the last word of science and of capitalist technique with the mass unity of the conscious workers engaged in big socialist production." Already this postulate has assured the possibility of overtaking and outstripping the level of technical and economic development in the advanced capitalist countries within the shortest period (in ten years at the utmost), at the same time "outstripping," by far, the level of material welfare for the toilers. This is assured, first of all, because the advantages of the uniform socialist plan, the absence of the obstacles of private property, etc., afford the possibility of a far more complete, universal and extensive utilization of the achievements of capitalist technique than is possible under capitalist conditions.

We have already seen last year how the Soviet farms and "machinery and tractor stations" have achieved a far more complete and rational utilization of the tractors than could be achieved by farmers and even by big agricultural companies in the United States.

As we shall presently see, the Soviet electric power stations, despite the relative backwardness of the equipment, have worked with a greater average load, resulting in a higher co-efficient of useful activity than has been the case in the capitalist countries.

The number of such facts is added to year by year, since the building of the new powerful combines (Ural Kuznetsk, etc.) should, upon the basis of the socialist plan, lead to an even more effective linking up of the branches of industry that are technologically allied. The task of the perspective plan is to investigate and to utilize to the utmost these advantages that do not exist under capitalism.

Of course, all this does not mean to say that the Soviet Union can be contented with copying the technique of the capitalist countries.

Already now we see how, for instance, in the domain of agriculture, the machinery, implements, and methods devised for small-scale capitalist farming cannot satisfy the requirements of large-scale socialist agriculture, so that we have to work for the creation of entirely new types of agricultural machines. The same situation arises in connection with the development of new construction in a number of other branches.

Hence, while relying upon the latest achievements of modern technique, the Soviet Union must, at the same time, in the drafting of the perspective plan, take full stock of the possibility and necessity of the conscious planning and directing of technical progress, of assigning " social orders " to inventors and designers of machinery upon a mass scale.

This problem is rendered very real by the tendencies of modern technical development. It is precisely in this way that the foundation for the new socialist technique ought to be laid upon the basis of the thorough electrification of production processes, of maximum mechanization and automatization of production, of a radical change in the type of labour, and in the role of the individual worker.

Another fundamental of technical reconstruction in the perspective plan is the emphasis put upon mass production in all branches of economy, especially as regards the mass production of the means of production.

Mass production, with all its peculiar technological features, was evolved during the last stage of capitalist development. But, on the one hand, it involved but a few countries (primarily the United States), and, on the other hand, but a few branches of production (automobiles, sewing machines, electric lamps, etc.).

Even in the United States the methods of mass production have not been extended to any considerable degree to the production of means of production, of machines, tools, etc. Moreover, the conditions of modern capitalism involve the placing of continually fresh obstacles to the development and further application of the methods of mass production, while frequently, even in branches where such methods are in use, their advantages are transformed by capitalism into disadvantages. Over-production, congested markets, and ominous depressions and crises have become the greatest obstacles to mass production. In the Soviet Union these obstacles do not exist, neither can there be any over-production, crisis, and unemployment. The requirements of the great masses of the toilers are growing at an unparalleled speed, calling for steady increase in the production of various commodities. The gigantic scope of industrial construction, and the beginning of the technical reconstruction of our agriculture upon an unparalleled scale, call for ever greater extension of the mass production of implements, metal, motive power, chemical products, etc. The emphasis on mass production calls, in the first place, for a maximum of standardization and interchangeability of parts, for specialization and co-operation between plants (" semi-manufacturing plants "), and for the adoption of the conveyor system of production in numerous branches of industry.

The fundamental, directing, and all-pervading principle in the perspective plan is unquestionably the PLAN FOR THE ELECTRIFICATION OF THE ENTIRE COUNTRY.

Ten years ago, when the first electrification plan of Soviet Russia was being drafted, Lenin wrote:

> " The only material basis for Socialism can be the large scale machine industry, capable of reorganizing agriculture also. Yet we should not limit ourselves to this general proposition. It is necessary to present it in concrete fashion. Large-scale industry, answering to the level of the latest technique, and capable of reorganizing agriculture, involves the electrification of the entire country."

1. PROBLEMS OF ELECTRIFICATION.

Already ten years have elapsed since the approval of the GOELRO—the State Electrification Plan. The basic economic indices of the GOELRO Plan have already been considerably surpassed. The electrical basis of the plan has been carried out with a substantial margin during the current year, except for a few of its backward sections (particularly as regards electrification of transport).

We are now confronted with the problem of drawing up the new perspective plan of electrification, which is to be the fundamental, decisive section of the entire general plan of development of the national economy of USSR, i.e., of the plan of developed Socialism.

Great, profound social and economic changes have taken place during these ten years.

All the branches of our socialist economy are showing a steady upward curve of development and expansion.

Already, last year, the socialised production relations affected the larger part of agriculture, directing the latter into the channel of the general economic plan, and creating the basis for a great technical revolution in the methods of agricultural production. The completion of the foundation of socialist economy will render the outlines of the fully-fledged socialist society even more definite and tangible and obvious to the masses of our people.

All these changes should find their reflection in the new perspective plan of electrification.

On the other hand, during these ten years there have occurred tremendous technical changes, particularly significant in the domain of electric power and electrification.

Technique does not stay still for a single day; not only does it progress, but it rushes forward, regardless, or, rather, in spite of the growing obstacles to its development under the conditions of decaying capitalism. In a way, these achievements of technique intensify the instability of modern capitalism. As was pointed out by Lenin, " The tremendously rapid growth of technique has as its concomitant even more of the elements of disproportion among the different branches of economy, resulting in chaotic conditions, in crises."

On the other hand, these successes of technique indicate the possibility, after the elimination of the capitalist obstacles, attaining an even more powerful, and, what is more, a planned, extensive, and universal utilization of the technical achievements for the development of the forces of production.

It stands to reason that the perspective plan of electrification must take stock also of those technical changes that have occurred during the past ten years, incorporating them in the scheme of future development.

What were the basic trends of these changes in the domain of electrification, and what possibilities do they open to the Soviet planned power production?

It will be expedient to examine these changes under three basic heads, representing the different branches (which, are, naturally, inter-connected), viz.: 1, The derivation of electrical power; 2, The transmission of current; and, 3, The utilization of electricity in various branches of economy. This subdivision corresponds on the scale of national economy as a whole, to the subdivision outlined by Marx as regards the analysis of a machine, namely: 1, The motor; 2, The transmission mechanism; and, 3, The working machine.

Electrification, in its full development (possible only under conditions of socialism), transforms the whole of the national economy into something like an uniform system of machines that is connected with the centralized electric power for the whole country, and with the uniform planning principle of the socialist state.

Let us now examine, in their essential features, the changes which have occurred in electrification during the last ten years, under the three aforesaid heads:—

I.

The basic source for the derivation of electrical power is still, and will unquestionably remain for the coming years, that of solid fuel, primarily coal, followed by lignite, peat, shales, etc. Let us note the tremendous changes which have occurred during these ten years as regards improving the methods of burning fuel.

It was in 1920, i.e., at the time when the GOELRO Plan was drafted, that in the United States the method of pulverized coal consumption was initiated upon a more or less extensive scale, having by then passed beyond the experimental stage. During these ten years the new method has rapidly gained supremacy at the largest electric power stations.

In the struggle against the new competitor, during these ten years we have also seen considerable improvements in the construction of various kinds of mechanical appliances for burning coal. To gain an idea of the extent of the mechanization of the processes of fuel consumption in the United States, it suffices to refer to the data of an investigation carried out in America in 1928 covering 98 per cent. of all plants having boiler installations.

According to these data, the proportionate fuel consumption was as follows:—

Pulverized Coal	17.0 per cent.
Overfeed Stokers	49.0 per cent.
Underfeed Stokers	3.0 per cent.
Chain-grate Stokers	28.7 per cent.
Hand Firing	2.3 per cent.
	100.0 per cent.

Thus, hand firing was retained only to the extent of 2.3 per cent. of the total coal burned (mostly existing in small, scattered plants), while 97.7 per cent. of the coal was burned by means of mechanical equipment of one kind or another.

In the large, modern electric power stations, the occupation of stoker is becoming a matter of the past, as though symbolizing the end of a distinct period, if not of a whole epoch in the development of technique.

There is yet another characteristic feature of these practical changes in the methods of the burning of fuel. About 1923 the utilization of pulverized coal, as well as of mechanical stokers of various kinds, seemed to have reached its utmost development, in view of the great loss of heat due to radiation. It was at that time that the so-called water-wall furnaces were introduced, which almost completely solved the problem of the radiation of heat, while at the same time affording a considerable increase in the capacity of boilers with the same heating surface (the walls being protected, the boiler becomes transformed into a sort of steam generator).

As a by-product of this change, there was also another, by no means negligible, consequence. In modern coal furnaces the temperature rises so high that the best grades of fire-bricks cannot withstand the heat. Whereas, 30 years ago, the brick walls of furnaces could withstand up to 12,000 hours of work, in 1920 the average durability was not more than 6—8,000 hours. The repairing of the brick walls became an important item of expense, apart from the great inconvenience involved, the interruption in working the boiler, etc. The introduction of metal walls, with water-cooling, at once solved this problem, enabling the working of 50,000 hours almost without repairs.

Another important change in the technique of steam power stations, is the rapid spread of the use of *high-pressure steam.* In American boilers a steam pressure of 100 and more atmospheres is quite normal. In some of the German plants even higher pressure is developed (up to 225 atmospheres). The possibility of this change was provided by the development of metallurgy, which began to turn out materials capable of withstanding such pressures and temperatures. Further development in this respect is retarded chiefly by this consideration, by the necessity of securing even more enduring metal compounds.

High pressure and high temperature, as is conclusively demonstrated by the practice of the last ten years, leads to a considerable increase in the capacity of boiler installations.

Despite the more complex equipment, the cost of investment per unit of production (steam) has been considerably reduced in the new boiler installations. Furthermore, high pressure opens the way to the wide development of the system of combined heat and electricity distribution. This development of recent years renders it possible in many cases to heighten the co-efficient of the useful service of electric power stations.

All these changes of the last ten years were accompanied by a tremendous INCREASE IN BOILER CAPACITY, so much so that in the United States there were gigantic boilers installed yielding as much as 1,200,000 lbs. of steam per hour. The furnaces of such boilers burn tons of coal in periods during which early 20th century boilers consumed only scuttlefuls (for instance, the furnace of a boiler at the new electric power station of Pittsburgh consumes 22 tons of coal per hour, while the entire boiler station, with a capacity of 60,000 k.w. is served by three workers).

The tendency towards larger boilers developed on parallel lines with the enlargement of steam turbines, creating a tendency towards the installation of power stations, so that ONE TURBINE IS SERVED BY A SINGLE POWERFUL BOILER. This achievement became possible only as a result of the tremendous growth in the reliability of all parts of power installations, rendering it possible to eliminate reserve boilers.

All these changes are most intimately inter-connected. Powerful gigantic boilers could not be working without mechanical stoking, since even a whole army of firemen armed with spades would not be able to cope with them. On the other hand, high pressures and high temperatures would be quite useless without the possibility of their application to modern turbines.

In this connection it is necessary to point out especially that during these ten years there have occurred tremendous technical improvements in the construction of steam turbines. It may be said that until quite lately there were no big revolutionary advances in the construction of turbines. But the modern turbine is the result of the accumulation of numberless small improvements, of the steady growth in the capacity of individual units, etc., rendering the modern turbine entirely unlike its predecessor of ten years ago.

It may be said, without exaggeration, that in the course of ten or twenty years the steam turbine has not only won complete mastery of the steam installations of electric power stations, but that it has also made greater progress than the steam engine has made from the time of Watt down to our own day.

Even more considerable has been the development of electric generators. As has been pointed out in an American journal, there has been a good deal of talk about the limits to the capacity of turbo-generators, while there is really no sign of such limitations.

By increasing the power of individual links, the capacity of turbo-generators has rapidly grown during these ten years, so that there is now a turbo-generator at an electric power station in Chicago developing 208,000 h.p., "the most gigantic and the most productive machine in the world."

All these changes taken together, and organically inter-connected, as already pointed out, have rendered it possible to build electric power stations generating upwards of half-a-million kilowatts (there is a station under construction for a million k.w.), i.e., of such high capacity that they become the foci of the economic life of entire industrial regions.

We cannot dwell here on the development of other types of motors (the development of internal combustion motors and of Diesel motors forms a separate chapter of technical changes), nor on the individual attempts to discover completely new methods in this sphere.

Let us only refer to a few instances :—

1. The mercury turbine, which has been successfully working for several years at one of the power installations in the State of Connecticut.

2. The German light-weight turbine of Wagner, which was built from light-weight metals and incorporated a number of other improvements. It weighs about two kilograms per h.p., i.e., about 30 times less than the usual turbine, while taking up far less space, so that it is a particularly desirable motor for transport (especially for ships).

3. The first attempts to employ internal combustion processes in turbines, and so on.

The technical progress of the last ten years, along the (so far) basic direction of transforming heat energy into electricity, as well as the researches along (so far) secondary ways, indicates that the whole sphere of technique is in a state of constant movement and change, that it is dynamic as never before.

In connection with this centralized process for the derivation of motive power, attention should be called to the increasing MECHANIZATION, and to the almost complete AUTOMATIZATION OF CONTROL. The almost complete automatization of control of these powerful aggregates supplying energy to whole districts and regions has become a quite real problem, a practical slogan for modern technique. You could see already two years ago in the United States huge electric power stations served by half a score of workers per shift, while the functions of those workers amounted almost exclusively to the watching of automatic registration gauges and to repair work. Since then even further progress has been made.

The same tendencies towards automatic control are even more applicable to *hydro-electrical* power stations. In the United States and Canada (especially in the outlying districts of the latter) during the last couple of years, a number of small and medium-sized hydro-power stations have been equipped with completely automatic or distant control. At the former the starting, synchronization, and stopping of the units is carried out automatically, depending on the demand for current, and the latter are controlled by employees located at a distance (sometimes many miles) from the plant, and watching the functioning of each part of the equipment by means of signals.

Already this cursory survey of some of the basic changes in technique in the derivation of electric power during the last ten years reveals the magnitude of the achievements in this domain. These achievements open up the possibility of securing, at an ever smaller expenditure of human labour, the motive power supplied by the huge natural sources of energy in its most perfect and docile form—electricity.

As we shall presently see, capitalism cannot utilise these possibilities, and is to some extent afraid of this development. Owing to the effects of the antagonisms of modern capitalism, this technical development proceeds very unevenly, by fits and starts, so creating a number of fresh antagonisms. These achievements constitute more or less individual records, while there are great numbers of backward power stations with obsolete equipment that are artificially maintained and supported by capitalist monopolies.

The Soviet Union, both in the absolute figures of output of electric current, and even more so as regards the consumption per capita, lags far behind the leading capitalist countries.

Nevertheless, the pace of the growth of electrification in the U.S.S.R. is several times quicker than that of the best periods in the United States. The absolute figures of the annual increase in power installations are also beginning to draw near to those of the United States, and are overtaking the European countries.

New capacity installations in the district power stations will reach in 1931 a total of about one million kilowatts (80 per cent. increase in one year), which is twice the capacity installation of Germany or of England (in 1928), in absolute figures, and dozens of times higher as regards the rate of increase.

The co-efficient of utilization of the Soviet electric power stations, as regards the number of hours worked annually, comprised in 1929-30 a total of 3,670 hours, or 42 per cent., as against 35.8 per cent. in the United States, 29.9 per cent. in Italy, 25.6 per cent. in Germany, 20.5 per cent. in England, and 20 per cent. in France. The result is that while, for instance, in England a new million of kilowatt capacity yields to the country an additional flow of energy of about 1.7 billion kilowatt hours, in the Soviet Union it provides no less than six billion kilowatt hours. The absence of capitalist obstacles, the Government's firm policy of electrification of the entire country, and the definite plans for electrical construction in the coming years, justify our confidence that all the aforesaid achievements in the technique of power derivation will find far fuller and wider application in the USSR than has been the case in the capitalist countries.

For this purpose it is essential, above all, to carry out extensive scientific research and to organise the proper prospecting of natural power resources. At the present time we have not yet investigated even a small portion of them. The number and the reserves of our coal basins, especially of Kuznetsk, are developing year by year. Quite lately there were discovered abundant coal reserves of splendid coking capacity, at Karaganda, in Kazakstan. The huge coal basins of Northern Siberia are almost entirely unprospected.

The brown coal of the Moscow district, lying quite close to the capital city, used to be considered worthless before the revolution. Now we know it to represent a valuable raw material for power production and chemistry, serving as the basis for a powerful industrial combine, distant gas supply, etc. And yet we have quite a number of such almost entirely unprospected brown coal basins. The peat reserves of USSR are incalculable. Yet it is only the complete mechanization of peat extraction (including preliminary work) and the wide development of the mechanical methods of peat extraction and burning that has rendered this fuel the basic source of power in a number of regions.

The oil shales of Middle Volga and of the Leningrad region are also a source both of cheap fuel and valuable raw material for the chemical industry.

Our petroleum output is so far limited to some five or six districts (Baku, Grozny, Maikop, Emba, Sakhalin). Yet there are many geological indications pointing to the presence of petroleum in a number of other districts. An obvious instance was furnished by the chance discovery of petroleum in the Urals.

We have almost failed to tap such valuable power resources as the natural gases, of which there are abundant emanations in Daghestan, in the southern parts of the Ukraine, in the Middle Volga Region, etc.

II

The technical changes in the transmission of electric power during these ten years have been no less considerable than those relating to the methods of derivation. In spite of the fact that the discovery of Deprès, the importance of which was immediately grasped by the farsighted mind of Engels, was made in 1891, nevertheless the development of high voltage transmission really made its first practical step only some ten or fifteen years ago. The inter-connection of electrical grids was in its embryonic stage as regards its scope and importance, involving but small and isolated districts. It was only in the post-war period that the first lines for the transmission of 120,000 volts were erected.

This makes us realize even more vividly the great insight of the genius of Lenin, who could see ahead, presaging the progress of technique, and during the years of great economic dilapidation, when but few factories were showing even feeble signs of existence, insisted upon the working out of a plan of electrification, based on the building of central power stations interconnected by high voltage transmission lines.

At the present time there is extensive use made in the United States of the transmission of 220,000 volt power. In many cases such installations are erected with a view to the eventual transmission of 400,000 volts.

The linking up of the electrical grids in the United States has advanced so far that all the main large stations, situated over the whole territory, from the Mississippi on the west to the Atlantic Ocean on the east, and from the Gulf of Mexico on the south, down to (and across) the Canadian border on the north, can exchange electrical power. This territory covers an area of about 900,000 square miles, i.e., twice that of Germany, France and England added together.

In a couple of years time this continuous network will be linked up with other systems west of the Mississippi, and then the grid of high voltage transmission lines will embrace a territory of one-and-half million square miles, with a population of about 100 million people, and a total load capacity of 20 million kilowatts will be possessed by the inter-connected power stations.

At the same time quite new perspectives and possibilities have lately been opened up as regards the transmission of electrical current.

In a report made by the chairman of the General Electric Company, Mr. Ryce, at the International Engineering Congress in Tokyo (1929), there were outlined possibilities, technically already feasible, for advancing the technique of power transmission in the near future. Laboratory experiments have already given a current of five million volts, enabling transformers, insulators, and other electrical apparatus to be submitted to tests on millions of volts.

At the same time, a series of successful researches have been made with respect to power transmission lines and the transmission of direct current of high voltage in particular. The possibility of such experiments has been brought about by the great development in late years of vacuum and gas-filled rectifiers of various kinds, and of analogous appliances for the re-conversion of direct into alternating current. These appliances (the so-called gasothrons) within a couple of years advanced from the laboratory stage into an important part of the equipment of the electrical industry. All this development indicates the possibility in the near future of alternating current of high voltage being converted by means of vacuum rectifiers into direct current of exceedingly high voltage. This direct current will be transmitted over long distances, and then by means of the thyrathron (one of the latest types of the aforesaid appliances) will be reconverted into alternating current. This renders possible the transmission of enormous volumes of electrical energy over distances of from 1,000 to 1,500 miles.

Among other things, this method will allow for an increase of two-three times in the capacity of the existing cable lines, which, of course, would be of tremendous economic importance. It is necessary to observe that the solution of these problems involves a complete revolution in the domain of insulating materials. The ways for this revolution have already been clearly outlined by the researches of the academician, Mr. Joffe.

To our perspective plan these changes in the domain of electrical transmission, partly already realized and partly rendered quite feasible, are of tremendous importance. They enable us to take up and elaborate in our perspective plan of electrification the problem that was repeatedly stressed by Lenin as the basic one, namely to accomplish the electrification of the ENTIRE country upon a uniform plan. This word ENTIRE was repeatedly stressed by Lenin in a number of his public speeches, articles, and letters.

This problem has by no means been fully tackled in the GOELRO plan.

The development of high voltage transmission in recent years affords the possibility of making fuller and more rational use of the huge resources of water power (the second and third stages of work on the Dnieprostroy, Volgostroy, Kama, Angara, the Niva and Kovda in the Murmansk region, Transcaucasia, the water power of the northern slopes of the Caucasian Ridge, the Tchirtchik, Tadjikistan, Altai, etc.) that are situated mostly on the borders of the country. It will enable us also to draft a plan for a network of thermic power stations of high capacity in the localities where fuel is obtained, particularly of cheap local fuel, with a view to centralizing the entire motive power of the country.

The general plan of electrification of the USSR should produce by means of a single network of high voltage lines a profound change in the distribution of the production forces in the country.

Mighty remote sources of energy will be directed into the common channel of motive power.

The " pioneer " stations will become the centres of quite new industrial regions that will be built from the very outset after the type of Combines, with the most rational utilization of the power, the raw material, the by-products of production, and so forth.

Cheap electric power will enable the organization of new technological processes upon unprecedented scales, particularly electro-chemical and electro-metallurgical, such as the production of aluminium, magnium, ferro-alloys, electric steel, and a number of chemical products.

A brilliant example of such a combine is presented by the Dnieper Combine, to be completed next year, upon the basis of the largest hydro-electrical station in Europe, being built on the Dnieper.

The general plan provides for the electrical unification of the industries of the Urals and the Kuznetsk Basin, embracing regions of colossal natural wealth that are 2,000 kilometers distant from each other. The unified network of transmission lines and of electrified railways should weld into one economic organism these great centres of coal deposits, metallurgy, chemistry, machine construction, non-ferrous metals, and large scale mechanized agriculture, that are unequalled anywhere in the world.

This is to be followed by the utilisation of the tremendous reserves of energy of the remotest rivers, such as the Angara, Yenissei, Amur, the streams flowing down from the Altai mountains and from the Pamir peaks in Tadjikistan.

The utilization of these electrical resources will permit such an extensive growth in the production of light metals as will ensure a complete revolution in machine construction and in transportation, setting up new speeds, bringing new victories to automatic control, high pressure, high temperature, and high voltage.

In the projects drawn up under the general plan, the production of electric power at the thermic stations is combined with the chemical treatment of coal, shales, and other kinds of fuel, with the burning of waste product, and also with an extensive plan of heat distribution to the surrounding districts, the centralized supply of steam and hot water for manufacturing purposes as well as for household use.

The general plan of electrification contemplates extensive electrification of the transport.

Finally, universal electrification will deal the last blow to the antagonism between town and village.

Thus, the unified network of high voltage transmission lines will create a gigantic power field in which the technique will move in the direction of the most advanced tendencies. The great revolution in modern physical science which capitalism cannot fully utilize will be realized throughout the vast territories of the Soviet Union.

In the preliminary drafts for the general plan of electrification of USSR the following basic stages for the construction of a unified network of high voltage lines are contemplated.

During the second Pyatiletka, i.e., by 1937, should be completed the interconnection of all the district power stations in the European part of the Soviet Union. The electric transmission lines (1) Leningrad-Moscow-Donetz-Basin-Azov-Sea-North-Caucasia; (2) Nijni Novgorod-Stalingrad; (3) Moscow-Nijni, and (4) Krivoi-Rog-Dnyeper-Donetz-Basin-Stalingrad, will form a single system of the Moscow-Nijni-Stalingrad-Donetz-Basin ring joined by the lines of : (1) Leningrad-Murmansk and (2) Transcaucasia, where the utilization of the power of mountain rivers and of regulating hydro stations of the mountain lakes (Sevan Lake, etc.) will furnish exceedingly favourable conditions for the complete electrification of Transcaucasia.

This basic network will be connected with that of the Urals, which links up the powerful stations of the metallurgical works, those of the Kizel coalpits, and the hydro stations of the Kama and Ural rivers.

Even before 1937 the European electrical network should be connected with the Kuznetsk Basin via Cheliabinsk, Kurgan, Omsk, and Novosibirsk. The Kuznetsk Basin line, by means of the stations on the Altai rivers and at the newly discovered Karaganda coal deposits, will be connected with the Central Asiatic Line, embracing en route the copper deposits of Kounrad and other districts of Kazakstan, which has barely begun to reveal its colossal natural wealth. The extension of these lines to the north will connect the Kuznetsk Basin with Minussinsk, where the water power resources of the Upper Yenissei are estimated at millions of kilowatts, and subsequently also with Krasnoyarsk, where the even greater power resources of the Middle Yenissei will become available.

The Central Asiatic electrical line should utilize the power resources of the rivers flowing from the peaks of Tyang-Shang and Pamir, the loftiest mountain ridges in the world. These resources are estimated at about 40 million kilowatts.

This will afford the possibility of the combined utilization of the rivers (in the first place, of the Tchirtchik, Naryn, and Vakhsh) both for electrification and irrigation purposes, which will permit the most extensive development of cotton growing. Moreover, there will be great consumers for the electric power, such as the chemical industry producing the necessary fertilizers for cotton growing, and non-ferrous metallurgy.

The second Pyatiletka contemplates only the preliminary steps so far (extensive prospecting and drafting) for the creation of the Angara Combine. The Angara will give a colossal concentration of water power (upwards of 10 million k.w.) regulated by the Baikal. Bearing in mind that in the proximity are situated huge coal basins, vast reserves of ores of iron, non-ferrous, light-weight, and rare metals, the largest timber areas of the world, etc., there can be no doubt that this region contains some of the largest centres of motive power and industry of the future.

Finally, the Far Eastern (Pacific) electrical grid will connect the power stations of the coal pits of the Vladivostok district with the hydro stations on the Amur (at the De Costri Bay) and its tributaries.

Thus, in the project of the general plan we see a series of regional transmission lines (European, Caucasian, Ural, West

Siberia, Central Asiatic, Angara, and Far Eastern) connected by inter-regional transmission lines.

Already in the course of the second Five-Year Plan, having reached the level of the most advanced capitalist countries as regards the total capacity of the electric power stations, the USSR will out-distance the latter as regards the production of electrical power, bearing in mind that already at the present time the number of working hours of the stations, and their degree of load, is ahead of those of America.

The general plan contemplates a further increase in the coefficient of exploitation of the electric power stations, bringing it up to a total of 6-7,000 hours annually.

As we have seen, the creation of a unified network of electrical transmission will render it possible to complete the radical reconstruction of the geographical distribution of industrial enterprises, so securing the maximum utilization of all the natural resources (especially local fuels of low calorific value). It will ensure the minimum loss of labour in passing from the derivation to the conversion of raw materials and to the manufacture of the finished products, the industrialization of the economically backward districts, and the thorough electrification of agriculture.

III.

The third sphere of technical changes in electrification during the last ten years embraces the METHODS OF UTILIZATION OF ELECTRIC POWER and the effect of these methods on the character of industrial processes. (In this respect we deal not in this report particularly with the problems of electro-metallurgy and electro-chemistry, i.e., of electrification applied to the derivation of new material and crude substances, as well as electrification of agriculture.)

The effect of electrification is first of all revealed in industry. Electricity in industry serves as a sort of universal lever for mechanization, automatization, and rationalization of production, completely transforming the conditions of industrial enterprises.

The degree of electrification in the industry of the advanced capitalist countries may serve as an extremely convincing indication of their technical level of industrial development.

By the commencement of 1930 the average level of electrification of industry was about 75% in the United States, about 70% in Germany, and only 50% in England.

The development of industrial electrification does not merely indicate quantitative growth. It inevitably leads, at a certain stage, to profound qualitative changes in the production processes. The electric motor in industry is not limited to the role of a convenient mechanical mover. Under the effect of the peculiar features of this motor, at a certain stage of electrification a change ensues in the very character of the work of the producing mechanism, in the aggregate working of the mechanism of a given plan, and in the connection between different plants.

Let us deal with the first question.

An interesting budget of material on the changes in this sphere is furnished in the collective report of the engineers of the Central German Association of Electrotechnical Industry (Berlin, 1930). The report bears the title of "Electric Motor Drives in Industry" ("Elektromotorische Antriebe in der Industrie" by W. Geyer, Prof. W. Philippi, and collaborators). The specialists state their dry facts, little suspecting the profound change in technical development indicated by these facts.

The report deals with four main periods of electrification of industry and mining.

The first period was one of more and more persistent attempts to apply electricity to industry. Electricity during this period was applied chiefly to tasks that could not be otherwise accomplished, or would present tremendous difficulties. Gradually the electric motor captured during this period the internal factory transportation, cranes, mine elevators, etc.

The second period is characterized chiefly by the concentration of the power resources of the enterprises. The electric motor penetrates into metallurgy, capturing the rolling mills. In the textile and some other branches, machines are introduced with separate motors, gradually leading to the elimination of transmission belts.

The third period is characterized by the extended use of electricity on precise and accurate processes of gauging, speed regulation, co-ordination of the movements of different parts of a machine or of different machines. (The authors overlook the most essential change which occurred during this period, namely, the introduction of the conveyor and of the continuous forms of production upon the basis of the electric motor. This, however, forms a separate subject.)

Finally, the fourth, present-day period is characterized by the increasing fusion of the electric motor with the manufacturing equipment, by the creation of an ELECTRICAL MACHINE in the true sense of the term. The electric motor is no longer an outsider as regards the manufacturing machine to which it supplies the motive power. It penetrates into the parts of the machine, becoming an inseparable part of it, constructively combined with it, directly affecting the character of the machine and of the production processes. Thus, the whole of the production process becomes as it were the function of the electric motor.

Taking the place of physical labour, the electric motor allows the achievement of such flexibility and precision as enables the carrying out of mechanical processes in such combinations as were quite impossible under the use of physical or steam power.

Thereby the electric motor opens to industry quite new possibilities of technical development, profoundly changing the whole character of modern production.

In the aforementioned report of the German electrical engineers we find characteristic examples of such changes in quite a number of industrial processes.

In the textile industry in recent years there was a development of electric spinning machines with mechanical changing of spools reaching 4,200 revolutions per minute, i.e., twice the speed of the mechanical spinning machine. The machine gives a great increase in productivity, particularly in the handling of rough fibres (hemp, sisal, etc.). The machine is started by a switch button, and the whole handling of it has been extremely simplified, having been reduced in some of the latest models to the manipulation of 3 buttons. The report mentions also some interesting experiments in the installation of small separate electric motors on some of the models of the spinning machines (making 10,000 revolutions per minute) on each spindle. This leads to the elimination of energy wasted on transmission without any considerable increase in the price of the machine.

Electric motors are of even greater importance in the production of ARTIFICIAL SILK, having exercised a tremendous influence upon the production processes of this young industry from its very inception. In late years, as the result of successful work by electro-technical specialists in this respect, a number of types of special electrical transmissions have been designed for machines handling artificial fibre.

Electrification already secures the automatic carrying out of the first chemical stage of the production processes. At the present time the development is in the direction of almost complete electro-automatical running in the manufacture of viscose, where the pressing of buttons controls the work of the whole series of units, with a total absence of belting, and a handful of people to attend to the factory. Attempts were also made to render automatic, by means of electricity, all the other production processes, such as the regulation of the flow and raining of liquids by means of electric valves, etc. In one central cabin are concentrated all the signalling, controlling, and gauging instruments, which kindle a sort of dynamic looking-glass picture of the course of all the production processes. By a turn of the "governor," or of a clockwork mechanism, a number of necessary processes are carried out automatically, as the mechanism is connected with the corresponding parts of the electrical appliances. Electrification secures a tremendous speeding-up of production. Thus, for instance, electrical centrifugal spindles give 10,000 revolutions per minute as compared with 5-6,000 revolutions on mechanical spindles. Lately, models have been introduced, giving as much as 20,000 revolutions per minute.

Extremely interesting is the effect of the electric motors on PAPER MILL machinery.

Here the development of the multiple motor drive has had a decisive effect upon the whole course of production, leading to a tremendous increase of productivity in recent years.

Beltings have almost completely disappeared in modern paper mills. The separate electric transmission has permitted the organisation of the whole course of production on consistent technological lines, the full utilization of space, ease of observation, and cleanliness of the premises. The even working of the motors has permitted the production of a fully homogenous product, the rapid measurement of the power spent and the production obtained. It has become possible to vary the speed of the processes at all their stages.

The paper-making machines have already been for a number of years the perfect example of continuous working, from handling the pulp to wrapping the dry rolls.

Lately it has been possible to extend the continuous belt principle also to the first stages of the process, beginning with the arrival of the pulp in the port or at the storehouse of the factory. By means of registering gauges it has become possible to control the whole progress of the production process. The multiple motor drive on the paper-making machines has allowed for the perfectly even and co-ordinated movement of the separate parts of these huge machines, with remote control of speed, pressure, etc. by means of electrical levers. On new models there are individual motors for separate rollers, drums, and other parts. The intricate electrical appliances for the synchronization of movements lead to great simplification of transmission and easier running of the machine. There is great simplification of control, which is mostly carried out by means of a series of buttons regulating the speed of the whole machine, the movement of the paper, the rise and drop of the rollers, etc. The cabin containing the gauges and press buttons may be located anywhere, constituting something like the brain of the machine from which one worker controls the continuous stream of production processes from the reception of the raw material to the turning out of the finished product. Here we see how the transformation of the machine into an "automatic factory," as described by Marx on the example of the paper machines of his time, upon the new electric basis, now assumes a quite new and more perfect form.

Similar effects of the new forms of electric transmission may be seen on the PRINTING MACHINES, especially on the powerful newspaper rotary machines of the so-called multiple type adapted for the most rapid change in newspaper circulation.

Such machines may be worked either in part or jointly as an integral whole, in any combination of parts and with considerable variation of speed. This adaptability is no hindrance to their huge capacity.

In view of the huge size of such machines, distant control is possible only by means of multi-motor electrical transmission. Each part of the machines has its dependent group of motors, regulating appliances, and levers, electrically interconnected. On this machine there is also a distant control, i.e., switch buttons on the switch boards, pressed in accordance with the indications of the gauging instruments.

Similar tendencies are developing more and more clearly in recent years on the various METAL WORKING LATHES.

The development proceeds primarily in the direction of supplying each lathe, and lately also individual parts of complex lathes, with separate motors. The tendency is towards the greatest simplification of the transmission between the motor and the working spindle, so as to curtail mechanical losses and simplify the control.

In some of the new lathes with multiple spindles, each spindle is equipped with a separate motor.

The curtailment of losses in transmission, and the avoidance of running the motor when the machine is not working, leads to an economy of about 50% in electrical power as compared with mechanical transmission.

Changes in the direction of the revolving part of the working lathe are also beginning to be effected by means of electrical transmission, but by switching over the electrical motor.

At the same time the apparatus is so arranged that the lathes can be operated without having to leave the place of work to start it. Here again the push button method is adopted both for switching on and off, as well as for regulating the number of revolutions, the direction, etc. All this renders possible and relatively simple the next step—automatic switching dependent directly upon the requirements of the working process. The simplest of such appliances are automatic fittings to turn off the current when work is finished. It also becomes easy to set the automatic appliance of the working machine at the most constant maximum capacity, for instance by changing the connection with the motor in accordance with the quantity of power required, or by changing the speed of working in accordance with the position of the article in hand, or of the tool, or by establishing a constant speed of cutting, and so on.

Needless to say, this development of individual electrical attendant motors, as compared with transmission belting, renders quite easy the inspection of the working premises and their adaptation to the conditions of the production process, facilitating the care of machinery, transportation, etc., and resulting in a tremendous increase of productivity. The starting and stopping of lathes is reduced to decimal fractions of a second. On revolving lathes, screw threading machines, automata, etc. when it is necessary to effect quick changes in the direction and speed of revolution, the switch control renders it possible to pass in less than a second from a speed of 1,500 revolutions per minute in one direction to a similar speed in the opposite direction. This is possible only with electricity. On large lathes and machines the electric motor drive means the elimination of jostling, of wear, of reduced capacity, unavoidable in mechanical transmission.

In some of the heavier machines the electrical motor drive makes its possible to dispense with flywheels, which leads to a great economy of power, weight and place.

In rolling mills the multimotor electrical belts afford also the possibility of almost complete automatization.

A worker in charge of a huge mill, standing at the little switchboard, performs only the initial function of the production process by turning on the motor in one direction or another. The further processes take place automatically. One worker can take care of a number of such switching appliances which are sometimes placed in a separate room, are easy to watch, and are isolated from the dust of the workshop. Thus, the basic control of the production process begins to be concentrated in a separate cabin equipped with gauges, signal lamps, and buttons. With such an equipment, the whole control of the huge rolling mill is carried on by two workers, one of whom operates the main motor and the adjustments of the rollers, while the other takes care of all the auxiliary appliances. Sometimes one worker watches at a distance the course of 7 or 8 production processes.

Very similar are the effects of the electric motor in the chemical industry. Electrification considerably simplifies the whole course of the production processes of chemical enterprises. The character of production makes it particularly expedient and profitable to apply remote control and to concentrate all the switches of the motors, i.e., the whole control over the production process in a central cabin isolated from the workshop, where the control is exercised by a couple of workers who watch the gauges and act by means of pushing one or another switch button. In some cases the working premises and the machinery located therein are only visited from time to time by a fitter to examine their condition and to repair any defects that may be found.

We are not going to dwell on other production processes. Everywhere the changes are quite analogous. Everywhere we find that " remote control " and " automatic control " upon an electrical basis are the technical slogans of the day.

Similar tendencies, so far in less pronounced form, may be observed also outside purely industrial enterprises. We find, for instance, similar attempts at remote and automatic control in some branches of mining activity, especially on surface mines of brown coal that are run on a mechanical basis, on the production of building materials, etc., on some hydro-technical works, etc.

Let us refer, even though cursorily, to the so-called Selsyn device for distant control which has been extensively applied in late years to a large variety of processes.

The first appliances of this type were installed for controlling the sluices of the Panama Canal. In the control cabin, situated several miles away from the sluices, a miniature reproduction of all the movements of the sluice locks is obtained by means of the Selsyn appliances. The man in charge can see at a glance, at a given moment, the position of all the numerous parts of the sluice, the level of the water, etc., as the Selsyn apparatus shows all the dynamics of the process. For instance, when the sluice gates are opened, there is a simultaneous opening of the miniature aluminium reproduction of the gates in the control room. There are also miniature chains reproducing all the movements of the chains which protect the sluices against damage by passing boats. An aluminium cylinder shows the level of the water with a precision up to 1-10th of an inch. In accordance with this minute reflection of the dynamics of the movement of the sluices obtained at a distance of several kilometers, the man in charge, by pressing a button, sets into motion at a distance the necessary mechanism by means of levers*. In late years the Selsyn mechanism has found wide application on automatic hydro-stations (for regulating the stream of water, for controlling the pressure on the pipes, etc.), for raising bridges at a distance, for controlling elevators, for theatrical illumination, for indicators of levels of liquids, gas pressure, etc. In fact, these appliances are constantly being put to new, sometimes quite unexpected tasks. In the near future the Selsyn mechanism will no doubt be applied to complete automatic control of underground electrical railways, etc.

Mechanisms of this type, in conjunction with electrical levers, render possible the adoption of automatic control in the most diverse industrial enterprises.

The foregoing material furnishes but the most fragmentary picture of the great scope of these changes. These changes affect all branches of industrial technique without exception.

American technical literature in the most diverse branches of industry during the last year has been full of notes about sundry

* It is necessary to observe that it is relatively easy to bring appliances of this type under wireless operation, and experiments in this direction are secretly carried on in the various War Offices.

inventions, improvements, new models in this direction. All this goes to show that technical development is on the eve of a most profound revolution in the methods of production.

Just as happened on the invention of the steam engine at the close of the 18th century, or in connection with the development of electrical technique in the middle of the 19th century, inventions, improvements, and new devices are carried out on a mass scale in different directions simultaneously with progress in the main direction.

This is now taking place upon a much larger scale. One of the consequences of this development is the following tendency. The motor, the transmission mechanism, and the working machine, under the influence of steam technique, were becoming more and more specialized and separated from one another. Upon the new electrical basis they are, on the one hand, becoming even more separated, separated sometimes for hundreds, and in the near future for thousands of miles. Yet, on the other hand, they are beginning to overcome this separation, being merged into one organic whole.

All these changes, when fully developed on a socialist basis, will have the following basic consequences for the worker. They will secure a tremendous growth of productivity with the maximum economy of labour, power, and materials.

This development opens the road for ever greater automatization in all branches of production. This automatization upon the basis of electricity will allow for the replacement of human labour, not only on hard physical work (completing in this respect the mechanization in its previous forms), but also on those monotonous, brain and body sagging, semi-skilled functions of the " operator " that are so characteristic for the semi-automatic mass production, the Ford type of conveyor production.

The type of labour of the worker at the switchboard of a modern electric power station presents already a characteristic example of the new type of labour in an automatic electrical enterprise.

Quite analogous forms will be taken (and partly have already been taken) by the labour of the worker on the most diverse mechanisms and in the most diverse production processes.

The control of metallurgical processes in blast furnaces, steel furnaces, rolling mills, sundry metal-working lathes, scores of automatic weaving looms, big paper mills, huge rotary printing presses, chemical apparatus, dispatch stations of electrified railways, etc., is becoming of a uniform type and character as regards the labour processes.

Already the present level of capitalist technique, already the manifest trend of present technical development indicates the quite definite possibility of securing to mankind huge supplies of electrical power, and of cheap power at that, to be distributed and utilized without any extensive expenditure of human labour.

This possibility of utilizing the huge power resources is the augury of the possibility and proximity of the period when " the forces of production will flow in a wide stream."

Nevertheless, under the conditions of modern capitalism, of capitalism which once called into being these achievements of technique, which has inaugurated this furious race of industrial progress, we find that at each step obstacles are raised to this development, that each success in technique intensifies the old, and creates new antagonisms.

Let us examine a few concrete instances of these antagonisms in relation to the electrification tendencies described above.

All the technical and economic conditions urge the enlargement of electric power stations and of their installations, such as boilers, turbines, and generators.

For instance, according to the detailed estimates by German engineers at the Second International Power Conference, the larger the size of turbo-generators the less the relative expenditure of capital, the less waste in heat, the lower the cost of maintenance of power stations, etc.

Working at the same load, the largest turbo-generator gives the lowest cost of electrical current.

Exactly the same deductions were made by American experts.

For instance, the projects of the Toronto station show that by having larger aggregates, with the same total capacity of the station, it would be possible to effect a 20% reduction in costs.

The same applies to boilers. On replacing a number of small boilers by a few large ones with high capacity and a high co-efficient of useful action, a tremendous saving is effected on all maintenance expenditure.

The increase in the general capacity of electric power stations means a considerable reduction in their relative space and area.

For instance, according to German data, the average area of a station per 1 kilowatt comprises, for stations of 1,000 k.w. capacity—0.14 sq. meters, and for stations of 150,000 k.w. only 0.06 sq. meters.

The corresponding average figures in cubic terms are:—2.2 meters and 1.6 cubic meters, repectively.

According to American data, the increase in the capacity of turbo-generators causes a considerable reduction in costs per unit of capacity, flooring space, cubic space of premises, etc.

The new types of turbines allow us, upon the same area and in the same premises, to install, instead of the old turbo-generators, new ones of 2-3 times the capacity.

Thus, Ford has recently replaced the old 25,000 k.w. turbo-generator by a new one of 110,000 k.w. The Jersey City station has replaced two old turbo-generators by three of the new type. It was reported by one of the American delegates at the second Power Conference that "minute analysis of costs has shown the maintenance cost of power stations per kilowatt to be sharply reduced by increasing the capacity of boilers and turbo-generators, and that so far there were no evident limits to this development."

Technically, it is already now possible to build aggregates of far greater capacity than those in existence. Formerly such large installations were not built, on account of possible damage, as, the sudden switching off of such a huge aggregate would cause a catastrophe. Now, with the wide interconnection of electrical lines, this danger is largely obviated, because even if the most powerful turbo-generators go out of commission, the electrical system has sufficient reserves to maintain the pressure. What is the matter, then? Why, bearing in mind the perfect rationality of this tendency of technical development, is it not applied on a large scale, but only in some isolated record cases?

First of all, the aforesaid advantages accrue only as a condition of working at full load. The less the habitual load, the smaller should be the aggregates.

The same as regards boilers. If not working at full load, a greater number of boilers is required than of turbo-generators, and consequently, smaller, technically less perfect boilers. In fact, one of the delegates at the Power Conference had to confess: "Economic causes are militating against technical expediency."

Of great importance also as regards the production costs of current is the co-efficient of load, i.e., the degree of utilisation of power stations.

Even according to pre-war data an aggregate of 1,000 h.p. required, at four hours work daily, twice as much coal per. k.w.h. as if working continuously. When the load is reduced, the co-efficient of useful action is reduced relatively even more rapidly.

Thus, reduced load does away at once with all the tremendous, quite unquestionable technical and economic advantages of big "mass" production of electrical current.

Yet the conditions of modern capitalism inevitably cause an incomplete load of the power stations at the present time, with a tendency to the further reduction of their utilisation.

That is why at the second International Power Conference nearly all the speeches and reports gave greatest prominence to the problem of load. That is why in the United States the most burning question in the sphere of electrification is the race to increase at all cost the co-efficient of load (especially of the so-called domestic use of current). Nevertheless, in spite of all efforts in this respect, the disproportion between the growing capacity, technical improvement and rationalization of the power stations on the one hand, and the degree of their utilization on the other hand, far from diminshing, has been steadily increasing in recent years.

The co-efficient of utilization of the electric power stations in 1929-30 was about 35% in the United States and France, about 25% in Germany, about 21% in England (in U.S.S.R. about 40%).

Such was the situation before the world economic crisis of 1930, which has made matters a good deal worse in this respect.

Prior to the crisis the production of electricity by the main capitalist countries, in absolute figures, was steadily growing, even if showing a distinct tendency towards a declining rate of growth.

In 1930 there was a distinct setback in a number of countries, a decline set in not only in the relative, but also in the absolute figures of electricity production. In the United States, in Germany, and in England, the indexes began to climb downwards like mercury in a thermometer exposed to the frost. This meant not only an immediate check to a number of the technical improvements described above, but also the transformation of these achievements into handicaps.

Another side to the same question, another antagonism of electrification under the conditions of modern capitalism, is raised by the development of monopolies. Under capitalist conditions the erection of big stations, the creation of central power stations, the interconnection of conducting lines, the tendency of development of the electro-technical industry, etc., leads to the emergence of particularly powerful capitalist monopolies that are closely intertwined with leading groups of financial capital.

The growth of these monopolies inevitably leads to particularly sharp manifestations of the tendency to decay.

Let us refer, for instance, to the declaration made by the American Ambassador to Germany at the 2nd International Power Conference, who declared that the electric power stations in the United States, which enjoy a monopoly, charge the consumers a price that is 15 times the actual cost.

Or we may refer to the bitter fight put up by the electrical trusts against a series of wholly feasible projects for hydro-stations (Muscle Shoals, St. Lawrence River, and others).

The electrical trusts are squandering millions of dollars in this fight, maintaining special lobbies at Washingon for the corruption of officials, sticking at no means to hinder the development of possible competitors.

This, in the United States, in the most powerful and full-blooded country of modern capitalism. Is it to be wondered at that in Germany one of the reporters at the International Power Conference declared that the modern conditions permit of interest only in "tested constructions" and advised his hearers to beware of all considerable technical innovations.

The exceptional acuteness and strain reached by these capitalist antagonisms in the realm of electrification are clearly demonstrated by the struggle developing in the United States in connection with the attempts of the electrical trusts to capture a complete monopoly of electricity supply.

Thus, for instance, here is a portion of a speech recently made by Governor Pinchot of Pennsylvania, on his official inauguration:—

> "Back of the public utilities in their attack on our American form of government, is the whole fabric of political corruption: the underworld, the protected racketeer and criminals of high and low degree."

This comes from a man entrusted with the high office of Governor of one of the greatest commonwealths of the United States, and it was delivered, not as an extempore speech, but as the carefully prepared inaugural address of that Governor.

The same was declared by Senator Norris in the United States Senate on May 9th, 1930. Referring to a statement made by President Hoover concerning the "great public utilities of the country," the Senator said:—"How can he refer to them as the 'great public utilities' in the face of the record of the Federal Trade Commission, where it is shown that for years they have been deceiving and robbing the American people? They have been engaged in politics from top to bottom in every State, dirty politics, disreputable politics, which ought to bring the blush of shame to every patriot. I do not like to have our President referring to them as 'great public utilities.' They have never done anything except to feather their own nests, and deceive the very people who, by their pennies, contribute to their wealth."

Senator Norris said further:—"The reason we do not have cheap power in this country is because of the faults and misleading propaganda which the power trust has circulated in every school district, in every municipality, among the churches, the lodges, the Boy Scouts, and women's clubs, poisoning the minds of our people, and especially of the rising generation, in the effort to create opposition to the kind of government control which has produced such cheap power in Ontario. That is what stands in the way of happiness and prosperity in millions of American homes to-day."

The electrical trusts took up the gauntlet. Thus, the Vice-president of Southern California Edison Company, in an article entitled, "Political Sabotage and the Power Industry" ("Electrical World," March 21st, 1931), refers to the above utterances as " malicious destruction of property under the guise of political activity in the public interest," as " sabotage, not by the workers, but by leading political parties." He asks for an indictment for the use of the " malignant tongue of slander," he urges a "crusade against this most powerful group of irresponsible agitators holding high places in politics." Of course, Pinchot, Norris, and the others cannot be in any way accused of subversive activity as regards private property, the American Constitution, etc. They simply represent the interests of those capitalists who are not directly connected with the electrical trusts and the financial groups supporting them, and whose profits are reduced by the high rates charged for electric power.

To us this furious fight between different capitalist groups is interesting only to the extent that it clearly demonstrates the obstacles raised by capitalist reality at every step to the planned development of electrification.

Very similar obstacles and antagonisms are referred to by a prominent German electro-technical specialist in the journal, " Technik der Wirtschaft " (1930).

He describes in detail how modern electrification creates a series of knotty, mostly unsolvable problems for capitalist law based upon private property. The laying of high voltage lines presupposes the consent of numerous owners of land property, etc. This causes terrible delay in the realization of many projects, as well as additional overhead expenditure. In the majority of cases it takes the shape of either the monopoly of one company in a given district, confirmed by law, and as the inevitable consequence, " legal " robbery by exorbitantly high charges for power, or, as is the case in Spain, a chaos of competing stations with parallel conduit lines, varying voltages, and a huge squandering of funds.

Still more complicated is the question of the very construction of large power stations, particularly water power, which require the planned regulation of the whole river system, i.e., the actual handing over to the concessionaire of jurisdiction over entire regions.

The writer feels bound to observe that under such conditions the dream about a general electrical, and even energetical plan, are merely literary exercises. " One cannot even think of such a curbing of free competition and of private property."

Still worse is the situation in England in this respect. We are told, for instance, in a report by Mr. H. Blankenhorn of the Academy of Political and Social Research, that " in England there are too many stations of the most varying capacity and age in one and the same district, while the voltage and the period numbers are various; there is a mass of legal limitations and conflicts, and the adoption of distance transmission is obstructed."

For instance, in the London district " there are 77 power stations, 50 nets, a couple of dozen of different kinds of voltage, and half a dozen different period numbers; the nets are intertwined, but the stations cannot lend any help to each other in case of a breakdown."

He goes on to observe that the power companies are trying not to cheapen power, but to insure their respective districts against competitors. As was stated in the House of Commons, " they are anxious to obtain and keep the monopoly while there is still time for it, and not to produce power."

Mr. Blankenhorn arrives at the conclusion that the technical problems of electrification, the creation of large power stations, unified nets, etc., " is obstructed in England by electrical appanages and by power parishes."

Lately desperate attempts have been made in England to rationalize even to some extent the supply of electric power, to electrify the railways, etc., but such attempts invariably meet with stout resistance on all sides.

Suffice it to mention that the electrification of railways in England, which can be carried out with the greatest ease and which promises tremendous economy in fuel, is being carried out exceedingly slowly because the railways companies fear the pressure of the coal companies which take up 30% of the freights carried by them.

Even in Japan, a country with a young industry and with a rapid development of electrification, one frequently finds upon one river the hydro installations of different companies and the parallel lines of competing transmission nets to the same centres of power consumption.

Such being the situation, it is no wonder that the high-sounding projects of Oliven, Schönholzer, Vieille, etc. for the carrying out of planned electrification throughout Europe remain a hopeless utopia, regardless of the fact that technically they are already quite easily realizable, and promise tremendous economic advantages. It is characteristic that these projects for an " electrical United States of Europe " are the clear reflection of the rivalry between the different capitalist states, each of them trying to get its own country to be the electrical heart of Europe, the central dispatcher of electrical current, controlling the whole net and deriving, on account of this, tremendous economic, financial, and strategic advantages. All these projects thus reflect the fundamental antagonisms of modern capitalism, and their realization within the bounds of capitalism is just as impossible as the elimination of these antagonisms.

Space will not permit us to dwell on innumerable other manifestations of decay in the sphere of electrification. At every step, upon every point of technique, we see with ever greater clarity how " the technique of capitalism day by day grows more and more beyond those social conditions which doom the toilers to wage slavery." (Lenin.)

As we have seen, already within the confines of capitalism there are now maturing the rudiments of tremendous technical changes in the sphere of electrification, of changes which ensure a veritable revolution in all the methods of production and in the character of labour.

Nevertheless, not only is capitalism incapable of widely and universally developing these tendencies, but it also fears their growth and development at the present stage, endeavouring in many cases to stifle such rudiments of the new technical revolution, to hinder their spreading and universal adoption, or it applies them in such a manner as to turn them against the working class, against the large masses of the toilers (especially against the small peasants), against the existence and development of the entire human race. Most characteristic in this respect is the fact that all the latest discoveries in the sphere of electrification, especially automatization of control, distance control, television, etc. are finding their primary application in the preparations for new imperialist wars. Frequently they are kept secret for years in the safes of War Offices, so as to prevent them from being applied before the outbreak of war. These technical discoveries, weapons for the mighty growth and development of the forces of production, are turned by modern capitalism into the most terrific weapons of destruction and annihilation of the forces of production, and in the first place, of the millions of toilers.

However, quite apart from the immediate war preparations, all the miracles of modern technique are utilized by capitalism, in the first place, as a weapon for the suppression of the working masses, for worsening their conditions of labour, for increasing the exploitation. As was written by Lenin as regards electrification of agriculture in France: "Under capitalism, electrification inevitably leads to increased PRESSURE BY THE BIG BANKERS both on the workers and ON THE PEASANTS. While retaining the power of the capitalists, electrification cannot be planned nor rapid, and if it is at all accomplished, it will mean a new bondage for the peasants, a new slavery for the peasants expropriated by the financial oligarchy."

The construction of socialism alone can develop these rudiments and tendencies of the new, maturing technical revolution upon the basis of electrification, making them the basis of all the processes of production, utilizing them for an unprecedented expansion of the forces of production, for the development of the new forms of labour, of the new worker, of the new man.

The entire picture of the modern capitalist forms of industrialization will have then become a strange, hideous reminiscence.

Already now we see the embryonic beginning of a regeneration of a number of ideas and imagery in this respect.

Thus, the poetry of factory chimneys is already now becoming a symbol of the senseless waste of fuel power.

Smoke, soot, filth, din, hitherto insolubly bound up with big factory production, will have passed into oblivion. It will have become unnecessary to gather large human masses in individual plants, since mass production will have become possible even with a relatively smaller number of employees. The control cabin, with the switchboard, signalling apparatus, sundry gauging instruments, and start buttons—such will be the labour atmosphere of the automatic factory of the future, an atmosphere that is clearly and definitely discernible already to-day, upon the basis of present achievements and unquestionable, quite real tendencies of further technical development. This development promises the greatest emancipation of human toil. In a number of languages the word " labour " is etymologically connected with the word " slave," and the word " toil " with the word " hardship." This connection will completely disappear in practice.

As we have seen, the beginnings of all these changes, the first heralds of new technical revolutions are already now knocking at every door in all branches of production, as mature, burning problems of technical developments, which the growing obstacles of modern capitalism prevent from spreading in a mighty stream.

In the U.S.S.R. these obstacles do not exist, and regardless of a number of difficulties incidental to the transition period, we must take stock of these changes and of their dynamics in the drafting of our perspective plan.

Here, more than ever before, it is essential " to look forward, in the direction of technical progress," as was done by Marx, Engels, Bebel, and Lenin. Thus, Lenin wrote about the problems of the GOELRO plan, at a period of the greatest difficulties, the war still raging, tremendous dilapidation of industry and of agriculture:—

> " It (the plan of electrification.—M.R.) should be now drawn up so as to PRESENT AN OBJECTIVE, POPULAR, CLEAR AND LUCID (FULLY SCIENTIFIC IN ITS BASIS) PERSPECTIVE TO THE MASSES. Let us set to work and in 10-20 years we shall electrify the whole of Russia, both industrially and agriculturally."

This task in its initial stage (the first ten years) has been fundamentally accomplished. The Soviet Union did attract the masses by the great programme of electrification, and then, of the Five-Year Plan. The very word, Pyatiletka, has now become international (like the word Soviet), inspiring the masses of the toilers in the U.S.S.R. to work for the further development of Socialist construction, for the surmounting of all obstacles in its path, for the complete remodelling of the technical basis of the national economy.

A basis of support in the struggle for " clear and lucid perspectives " is furnished by the movements of the workers, masses who have created their Socialist emulation and shock work, and the mass struggle for rural collectivization.

It is this movement of the millions which enables us now, 10 years after the adoption of the GOELRO plan and the commencement of the reconstruction period of our economy, to complete in 1931 the foundation of Socialist economy, to assure the fulfilment of the Pyatiletka in 4 years, and in many branches even within a shorter period.

Unlike other building construction, while completing the foundation of Socialist economy we do not yet posses, neither can we possess, exactly detailed plans of the edifice which is beginning to grow up on this foundation.

Nevertheless, separate contours of the new structure are becoming more and more clearly discernible. standing out in ever closer and clearer relief. And we can already give fillip to these contours in the new perspective plan for the further development of Socialist construction, in the 2nd Five-Year Plan. And in that section of this plan, which should furnish the fundamental characteristics of the technical basis of construction we can already, relying on the one hand upon the social and class changes of recent years, and on the other hand upon a study of the trend of development of modern technique, and of the technical changes which became discernible after the adoption of the GOELRO plan, draw up a new " clear and lucid (fully scientific in its basis) perspective " of Socialistic production and labour upon the basis of universal electrification.

THE SOCIAL AND ECONOMIC ROOTS OF NEWTON'S 'PRINCIPIA.'

By B. HESSEN.

THE SOCIAL AND ECONOMIC ROOTS OF NEWTON'S 'PRINCIPIA.'

By Prof. B. HESSEN.

CONTENTS.

INTRODUCTION:

MARX'S THEORY OF THE HISTORICAL PROCESS.

The work and also the personality of Newton have attracted the attention of scientists of all ages and nations. The enormous extent of his scientific discoveries, the significance of his work to all the later developments of physics and technology, the notable exactitude of his laws justifiably arouse special respect for his genius.

What placed Newton at the turning-point of the development of science and gave him the possibility of indicating the new roads of this progressive movement?

Where are we to seek the source of Newton's creative genius? What determined the content and the direction of his activities?

These are the questions which inevitably confront the investigator who takes as his task not the simple assembly of materials relating to Newton, but who wishes to penetrate into the very essence of his creative work.

"Nature and nature's laws lay hid in night;
God said 'Let Newton be!' and all was light."

Said Pope, in a well-known couplet.

Our new culture, declares Professor Whitehead, a famous British mathematician, in a recent book: "Science and Civilisation" owes its development to the fact that Newton was born in the very year of Galileo's death. Only think what the history of the development of humanity would have been if these two men had not appeared in the world.

The well-known English historian of Science, F. S. Marvin, a member of the presidium of this International Congress, associated himself with this view in his article: "The meaning of the 17th Century," which appeared a couple of months ago in "Nature."

Thus the phenomenon of Newton is regarded as due to the kindness of divine providence, and the mighty impulse which his work gave to the development of science and technology is regarded as the result of his personal genius.

In this lecture we present a radically different conception of Newton and his work.

The quotations made in this essay have been translated from Russian. The chief exceptions are the quotations from "Nature" in Chapter 5.

Our task will consist in applying the method of dialectical materialism and the conception of this historical process which Marx created, to an analysis of the genesis and development of Newton's work in connection with the period in which he lived and worked.

We give a brief exposition of the basic assumptions put forward by Marx which will be the guiding assumptions of our lecture.

Marx expounded his history of the historical process in the preface to the "Critique of Political Economy" and in the "German Ideology." We shall attempt to give the essence of the Marxian viewpoint as far as possible in his own words.

Society exists and develops as an organic whole. In order to ensure that existence and development society must develop production. In social production people enter into definite inter-relationships which are independent of their own will. At every given stage these relationships correspond to the development of the material productive forces.

The aggregation of these productive forces forms the economic structure, the real basis on which the juridical and political superstructures are raised.

The definite forms of social consciousness also correspond to this basis.

The method of production of material existence conditions the social, political and intellectual process of the life of society.

It is not the consciousness of human beings which determines their existence, but on the contrary their social existence determines their consciousness. At a certain stage of its development the material productive forces of society come into antagonism with the existing production relationships, or with the property relationships within which they have hitherto developed (which is only a juridical expression of the same thing).

From being forms of productive forces they are transformed into fetters of those forces. Then follows the period of social revolutions. With the change of bases there occurs a transformation in all the enormous super-structure also.

The prevailing consciousness during these periods has to be explained by reference to the antagonisms of material existence, to the existing conflict between the productive forces and the production relationships.

Lenin remarks that this conception of the materialistic interpretation of history eliminates two chief defects of the previous historical theories.

Previous historical theories considered only the intellectual motives of the historical activity of people as such. Consequently they could not reveal the true roots of those motives, and consequently history was justified by the individual intellectual impulses of human beings. Thus the road was closed to any recognition of the objective laws of the historical process. "Opinion governed the world." The course of history depended on the talents and the personal impulses of man. Personality was the creator of history.

Professor Whitehead's above-quoted view of Newton is a typical example of this limited understanding of the historical process.

The second defect which Marx's theory eliminates is that the subject of history is not the mass of the population, but the personalities of genius. The most obvious representative of this view is Carlyle, for whom history was the story of great men.

The achievements of history are only the realisation of the thoughts of great men. The genius of the heroes is not the product of material conditions, but on the contrary the creative force of genius transforms those conditions, itself not being in need of any extrinsic material factors.

In contradistinction to this view Marx observes the movement of the history-making masses and studies the social conditions of the life of the masses and the modifications in those conditions.

Marxism, as Lenin emphasises, pointed the way to an all-embracing study of the process of origination, development and decline of social systems. It explains this process by considering all the aggregation of contradictory tendencies, reducing them to the exactly determined conditions of existence and the production of the various classes.

Marxism eliminates subjectivism and arbitrariness in the selection of the various "dominating" ideas or in their interpretation, attributing the roots of all ideas without exception, to the state of the material productive forces.

In class society the ruling class subjects the productive forces to itself and, by virtue of its domination of material force subjects all other classes to its interests.

The ideas of the ruling class in every historical period are the ruling ideas, and the ruling class distinguishes its ideas from all previous ideas by putting them forward as eternal truths. It wishes to reign eternally and bases the inviolability of its rule on the eternal quality of its ideas.

In capitalist society a separation of the dominating ideas from the production relationships occurs, and thus is created the view that the material structure is determined by ideas.

Practice has not to be explained by reference to ideas, but on the contrary the formation of ideas has to be explained by reference to material practice.

Only the proletariat, which has as its objective the creation of the classless society, is freed of limitations to its conception of the historical process and creates a true, genuine history of nature and of society.

The period during which Newton's activity was at its height corresponds with the period of the English Civil War and Commonwealth.

The Marxist's analysis of Newton's activity, made on the basis of the foregoing assumptions, will consist first and foremost in understanding Newton, his work and his world outlook as the product of this period.

THE ECONOMICS, PHYSICS AND TECHNOLOGY OF NEWTON'S PERIOD.

The general symptom of that section of world history which has come to be known as mediæval and modern history is first and foremost that during this period we have the rule of private property.

All the social and economic formations of this period preserve this basic symptom.

Consequently Marx regarded this period of the history of humanity as the history of the development of forms of private property, and distinguishes three subsidiary periods within the larger epoch.

The first period is that of the rule of feudalism. The second period begins with the disintegration of the feudal system and is characterised by the emergence and development of merchant capital and manufacture.

The third period in the history of the development of private property is that of the rule of industrial capitalism. It gives birth to large-scale industry, the application of the forces of nature to industrial purposes, mechanisation and the most detailed division of labour.

The brilliant successes of natural science during the sixteenth and seventeenth centuries were conditioned by the disintegration of the feudal economy, the development of merchant capital, of international maritime relationships and of heavy (mining) industry.

During the first centuries of the mediæval economy, not only feudal but to a considerable extent urban economy also was based upon personal consumption.

Production for the purpose of exchange was only then emerging. Hence the limited nature of exchange and of the market, the self-centred and static forms of production, the local isolation from the external world, the purely local connections of producers; the feudal estates and the commune in the country, the guild in the towns.

In the towns capital was in kind, directly bound up with the labour of the owner and inseparable from him. This was *corporation* capital.

In the mediæval towns there was no division of labour among the various crafts nor within those crafts among the various workers.

The insignificance of intercourse, the shortage of population and the limited extent of consumption hindered any further growth in the division of labour.

The next step in the realm of the division of labour was the separation of production from the forms of intercourse and the formation of a special class of merchants.

The bounds of commerce were widened. Towns entered into relationships with one another. There arose the necessity for the roads to be publicly safe, and the demand for good roads of communications and means of transport.

The newly developing associations between towns led to the distribution of production among them. Each developed a special sphere of production.

Thus the disintegration of feudal economy led to the second period in the history of the development of private property, to the rule of merchant capital and manufacture.

The emergence of manufacture was the immediate consequence of the division of labour among various towns.

Together with manufacture the relationships between the worker and the employer are modified. The monetary relationship between the capitalist and the worker makes its appearance.

Finally, the patriarchal relationships between the masters and the foremen are shattered.

Trade and manufacture created the great bourgeoisie. The petty bourgeoisie were concentrated in trades and were compelled in the towns to yield to the hegemony of the merchants and the manufacturers.

This period dates from the middle of the seventeenth century and continues to the end of the eighteenth.

Such is the schematic outline of the course of development from feudalism to merchant capital and manufacture.

Newton's activities fall within the second period in the history of the development of private property.

Consequently we investigate first and foremost the historical demands imposed by the emergence of merchant capital and of its development.

Then we consider what technical problems the newly developing economy raised for solution and we investigate to what grouping of physical problems and of science necessary to the solution of those problems these technical problems led.

We direct our survey to three outstanding spheres which were of decisive importance to the social and economic system we are investigating. These spheres are ways and means of communication, industry, and military affairs.

Ways of Communication.

By the beginning of the middle ages trade had already achieved considerable development. Nevertheless, the land ways of communication were in a very miserable state. The roads were so narrow that even two horses could not pass. The ideal road was one on which three horses could travel side-by-side, where, in the expression of the time (14th century) "A bride could ride by without touching the funeral cart."

Commonly, commodities were carried in packs. Road construction was almost non-existent. The self-centred nature of feudal economy gave no impulse whatever to the development of road construction. On the contrary, both the feudal barons and the inhabitants of places through which commercial transport passed were interested in maintaining the poor condition of the roads, because they had the right of ownership to anything which fell on to their land from the cart or pack.

The speed of land transport in the fourteenth century did not exceed five to seven miles in the day.

Naturally maritime and water transport played a great part, both in consequence of the great load-capacity of the vessels and also of the greater speed of transit: the largest of two-wheeled carts drawn by ten to twelve oxen hardly carried two tons of goods, whereas an average sized vessel carried upwards of 600 tons. During the fourteenth century the journey from Constantinople to Venice took three times as long by land as by sea.

Nevertheless even the sea transport of this period was very imperfect: as sound methods of establishing the ship's position in the open sea had not yet been invented, they sailed close to the shores, which greatly retarded the speed of transit.

Although the first mention of the mariner's compass in the Arabian book "The Merchant's Treasury" dates to 1242, it came into universal use not earlier than the second half of the sixteenth century. Geographical maritime maps made their appearance about the same time.

But the compass and charts can be rationally exploited only when there is knowledge of methods of establishing the ship's position,—i.e., when the latitude and longitude can be determined.

The development of merchant capital broke down the isolation of the town and the village commune, extended the geographical horizon to an extraordinary extent, and considerably accelerated the tempo of existence. It had need of convenient ways of communication, more perfect means of communication, a more exact measurement of time, especially in connection with the continually accelerating rate of exchange, and exact application of accounting and measuring.

Particular attention was directed to water transport: to maritime transport as a means of linking up various countries and to river transport as an internal link.

The development of river transport was also assisted by the fact that in antiquity waterways were the most convenient and most investigated, and the natural growth of the towns was linked up with the system of river communications. Transport over the rivers was three times as cheap as haulage transport.

The construction of canals also developed as a complementary means of internal transport and in order to link up the maritime transport with the internal river system.

Thus the development of merchant capital set transport the following technical problems:—

In the realm of water transport.

1. An increase in the tonnage capacity of vessels and in their speed.

2. An improvement in the vessels' floating qualities: their reliability, sea-worthiness, their lesser tendency to rock, response to direction and ease of manœuvring, which was especially important for war-vessels.

3. Convenient and reliable means of determining position at sea. Means of determining the latitude and longitude, magnetic deviation, times of tides.

4. The perfecting of the internal waterways and their linking up with the sea; the construction of canals and locks.

Let us consider what physical prerequisites are necessary in order to resolve these technical problems.

1. In order to increase the tonnage capacity of vessels it is necessary to know the fundamental laws governing bodies floating in liquids, since in order to estimate tonnage capacity it is necessary to know the method of estimating a vessel's water displacement. These are problems of hydrostatics.

2. In order to improve the floating qualities of a vessel it is necessary to know the laws governing the movement of bodies in liquids—this is an aspect of the laws governing the

movement of bodies in a resistant medium—one of the basic tasks of hydrodynamics.

The problem of a vessel's stability when rocking is one of the basic tasks of the mechanics of material points.

The problem of determining the latitude consists in the observation of heavenly bodies and its solution depends on the existence of optical instruments and a knowledge of the chart of the heavenly bodies and of their movement—of the mechanics of the heavens.

The problem of determining longitude can be most conveniently and simply solved with the aid of a chronometer. But as the chronometer was invented only in the thirties of the eighteenth century after the work of Huygens, in order to determine the longitude recourse was made to measurement of the distance between the moon and the fixed stars.

This method, put forward in 1498 by Amerigo Vespucci, demands an exact knowledge of the anomalies in the moon's movement and constitutes one of the most complicated tasks of the mechanics of the heavens. The determination of the times of the tides in dependence on the locality and on the position of the moon demands a knowledge of the theory of attraction, which also is a task of mechanics.

How important this task was is evident from the circumstance that long before Newton gave the world his general theory of tides on the basis of the theory of gravity, in 1590, Stevin drew up tables in which was shown the time of the tides in any given place in dependence on the position of the moon.

4. The construction of canals and locks demands a knowledge of the basic laws of hydrostatics, the laws governing the efflux of liquids, since it is necessary to know how to estimate the pressure of water and the speed of its efflux. In 1598 Stevin was occupied with the problem of the pressure of water and he saw that water could exert a pressure on the bottom of a vessel greater than its weight; in 1642 Castelli published a special treatise on the movement of water in canals of various sections. In 1646 Torricelli was working on the theory of efflux of fluids.

As we see, the problems of canal and lock construction also bring us to the tasks of mechanics (hydrostatics and hydrodynamics).

Industry.

Already by the end of the middle ages (14th and 15th centuries) the mining industry was developing into a large industry. The mining of gold and silver in connection with the development of currency circulation was stimulated by the growth of exchange. The discovery of America was chiefly due to the gold famine, since European industry, which had developed so powerfully during the 14th and 15th centuries, and correspondingly European commerce, demanded larger supplies of the means of exchange; on the other hand the need for gold forced especial attention to be turned to the exploitation of mines and other sources of gold and silver.

The powerful development of the war industry, which had made enormous advances from the time of the invention of firearms and the introduction of heavy artillery, stimulated the exploitation of iron and copper mines to a tremendous extent. By 1350 firearms had become the customary weapon of the armies of eastern, southern and central Europe.

In the fifteenth century heavy artillery had reached a high level of perfection. In the 16th and 17th centuries the war industry made enormous demands upon the metallurgical industry. In the months of March and April 1652 alone, Cromwell required 335 cannon, and in December a further 1,500 guns of an aggregate weight of 2,230 tons, with 117,000 balls and 5,000 hand bombs in addition.

Consequently it is clear why the problem of the most effective exploitation of mines became a matter of prime importance.

First and foremost arises the problem set by the depth at which the ores lie. But the deeper the mines, the more difficult and dangerous work in them becomes.

A quantity of equipment for the pumping of water, the ventilation of the mines, and the raising of the ore to the surface becomes necessary. In addition a knowledge of the sound opening up of mines and of the plan of their workings is necessary.

By the beginning of the 16th century mining had reached a considerable development. Agricola left a detailed encyclopædia of mining from which one can see how much technical equipment had come to be applied in mining.

In order to raise the ore and to pump out water pumps and lifting equipment (windlasses and horizontal worms) were constructed; the energy of animals, the wind and falling water were all put into service. A complete pumping system began to exist, since with the deepening of the mines the problem of removing the water becomes one of the most important of the technical tasks.

In his book Agricola describes three kinds of instruments for drawing away water, seven kinds of pumps, and six kinds of equipment for drawing off water by ladling or bucketing, altogether sixteen kinds of water-raising machines.

The development of mining involved enormous equipment for the working up of the ore. Here we meet with smelting furnaces, stamping mills, and machinery for dividing metals.

By the 16th century the mining industry had become a complex organism demanding considerable knowledge in its organisation and direction. Consequently the mining industry at once develops as a large-scale industry, free of the craft system, and so not subject to craft stagnation. It was technically the most progressive and engendered the most revolutionary elements of the working class during the middle ages, i.e., the miners.

The cutting of galleries demands considerable knowledge of geometry and trigonometry. By the 15th century scientific engineers were working in the mines.

Thus the development of exchange and of the war industry set the mining industry the following technical problems:—

1. The raising of ores from considerable depths.

2. Methods of ventilating the mines.

3. The pumping out of water and water-conducting equipment, the problem of the pump.

4. The transfer from the crude, damp-blast method of production predominant until the 15th century, to the more perfect form of blast-furnace production, in which the problem of air-blast equipment is raised, as it is in ventilation also.

5. The working up of the ores with the aid of rolling and cutting machinery.

Let us consider the problems of physics lying at the bases of these technical tasks.

1. The raising of ore and the task of equipping the raising machinery is a matter of arrangement of windlasses and blocks, i.e., of a variety of simple mechanical machines.

2. Ventilation equipment demands a study of draughts, i.e., it is a matter of aerostatics, which in turn is part of the task of statics.

3. The pumping of water from the mines and the equipment of pumps, especially of piston pumps, necessitates considerable investigation in the realm of hydro- and aerostatics.

Consequently Torricelli, Herique, and Pascal occupied themselves with the problem of raising liquids in tubes and with atmospheric pressure.

4. The transfer to the blast-furnace production at once evoked the phenomenon of great blast-furnaces with the necessary buildings, water-wheels, bellows, rolling machines and heavy hammers.

The problems of hydrostatics and dynamics set by the erection of water-wheels, the problem of air-bellows as also that of forced air for ventilation purposes also demand a study of the movement of air and its compression.

5. As in the case of other equipment, the construction of presses and heavy hammers brought into motion by utilising the force of falling water (or animal power) demands a complicated planning of cogged wheels and transmission mechanism, which also is essentially a task of mechanics. In the mill develops the science of friction and the mathematical arrangement of cogged transmission wheels.

Thus, leaving out of account the great demands which the mining and metal-working industries of this period made on chemistry, all the aggregate of tasks of physics fell within the limits of mechanics.

War and War Industry.

The history of war, Marx wrote to Engels in 1857, allows us more and more clearly to confirm the accuracy of our views on the connections between productive forces and social relationships.

Altogether the army is very important to economic development. It was in warfare that the craft order of corporations of artisans first originated. Here also we first find the application of machinery on a large scale.

Even the special value of metals and their role as currency were evidently based on their war significance.

So also the division of labour within various spheres of industry was first introduced in the army. Here in a tabloid form we find the entire history of the bourgeois system.

From the time of the application of gunpowder in Europe (it was used in China even before our era), a swift increase of firearms sets in.

Heavy artillery first appeared in 1280, during the siege of Cordova by the Arabs. In the 14th century firearms passed from the Arabs to the Spaniards. In 1308 Ferdinand IV. took Gibraltar with the aid of cannon.

The first heavy guns were extremely unwieldy and they could only be transported in sections. Even weapons of small calibre were very heavy, since no proportion whatever had been established between the weight of the weapon and the ball and between the weight of the ball and the charge.

Nevertheless firearms were used not only in sieges, but on war-vessels. In 1386 the English captured two war-vessels armed with cannon.

A considerable improvement in artillery took place during the 15th century. Stone balls were replaced by iron. Cannon were cast solidly from iron and copper. Gun-carriages were improved and transport made great strides forward. The rate of fire was accelerated. To this factor is due the success of Charles VIII. in Italy.

In the battle of Fornova the French fired more shots in one hour that the Italians fired in a day.

Machiavelli wrote his "Art of War" specially in order to demonstrate means of resisting artillery by the artificial disposition of infantry and cavalry.

But of course the Italians were not satisfied with this alone, and they developed their own war industry. In Galileo's time the Arsenal at Florence had attained to considerable development.

Francis I. formed artillery into a separate unit and his artillery shattered the hitherto undefeated Swiss pikes.

The first theoretical works on ballistics and artillery date from the 16th century. In 1537 Tartaglia endeavoured to determine the trajectory of the flight of a shot and established that the angle of 45 degrees allows the greatest distance to flight. He also drew up tables for directing aim.

Vanucci Biringuccio studied the process of casting and in 1540 he introduced considerable improvements in the production of weapons.

Hartmann invented a scale of calibres, by means of which each section of the gun could be measured in relation to the aperture, which gave a certain standard in the production of guns and opened the way for the introduction of fixed theoretical principles and empirical laws of firing.

In 1690 the first artillery school was opened in France.

In 1697 San-Remi published the first complete primer of artillery.

Towards the end of the 17th century in all countries artillery lost its mediæval, craft character and was included as a component part of the army.

Consequently experiments on the inter-relationship of calibre and charge, the relationship of calibre to weight and length of barrel, on the phenomenon of recoil, developed on a large scale.

The progress of ballistics went hand in hand with the work of the most prominent of the physicists.

Galileo gave the world the theory of the parabolic trajectory of a ball; Torricelli, Newton, Bernouilli and Enler engaged in the investigation of the flight of a ball through the air, studied the resistance of the air and the causes of declination.

The development of artillery led in turn to a revolution in the construction of fortifications and fortresses, and this made enormous demands upon the engineering art.

The new form of defensive works (earthwork, fortresses) almost paralysed the activity of artillery in the middle of the 17th century, and this in turn gave a mighty impulse to its further development.

The development of the art of war raised the following technical problems:—

Intrinsic ballistics.

1. Study of the processes which occur in a firearm when fired and their improvement.
2. The stability combined with least weight of the firearm.
3. Adaptation to suitable and good aim.

Extrinsic ballistics.

4. The trajectory of a ball through a vacuum.
5. The trajectory of a ball through the air.
6. The dependence of air resistance upon the flight of the ball.
7. The deviation of a ball from its trajectory.

The physical bases of these problems.

1. Study of the processes which occur in the firearm demands study of the compression and extension of gases—in its basis a task of mechanics, and also study of the phenomenon of recoil (the law of action and counter-action).

2. The stability of a firearm raises the problem of studying the resistance of materials and of testing their durability. This problem, which also has great importance for the art of construction in the given stage of development, is resolved by purely mechanical means. Galileo gives considerable attention to the problem in his "Mathematical Demonstrations."

3. The problem of a ball's trajectory through a vacuum consists in resolving the task of the free fall of a body under the influence of gravity and the conjuncture of its progressive movement with its free fall. Naturally therefore Galileo gave much attention to the problem of the free fall of bodies. How far his work was connected with the interests of the artillery and ballistics can be judged if only from the fact that he begins his "Mathematical Demonstrations" with an address to the Florentines, in which he praises the activity of the arsenal at Florence and points out that the work of this arsenal provides a rich material for scientific study.

4. The flight of a ball through the air is part of the problem of the movement of bodies through a resistant

medium and of the dependence of that resistance upon the speed of the movement.

5. The deviation of the ball from the estimated trajectory can occur in consequence of a change in the initial speed of the ball, a change in the density of the atmosphere, or through the influence of the rotation of the earth. All these are purely mechanical problems.

6. Accurate tables governing aim can be drawn up provided the problem of extrinsic ballistics is resolved and the general theory of a ball's trajectory through a resistant medium is given.

Thus we see that if the process of the actual production of the firearm and the ball, which is a problem of metallurgy, be left out of account, the chief problems raised by the artillery of this period were problems of mechanics.

Now let us systematically consider the problems of physics raised by the development of transport, industry and mining.

First and foremost we have to note that all of them are purely problems of mechanics.

We analyse in a very general way the basic themes of research in physics during the period in which merchant capital was becoming the predominant economic force and manufacture began to develop, i.e., the period from the beginning of the 16th to the second half of the 17th century.

We do not include Newton's works on physics, since they will be subjected to a special analysis. A comparison of the basic themes of physics enables us to determine the basic tendency of the interests of physics during the period immediately preceding Newton and contemporary with him.

1. The problem of simple machines, sloping surfaces and general problems of statics were studied by: Leonardo da Vinci (end of 16th century); Ubaldi (1577); Galileo (1589-1609); Cardan (middle of 16th century); and Stevin (1587).

2. The free fall of bodies and the trajectory of thrown bodies were studied by: Tartaglia (thirties of the 16th century); Benedetti (1587); Piccolomini (1598); Galileo (1589-1609); Riccioli (1652); The Academy del Cimente (1649).

3. The laws of hydro- and aerostatics, and atmospheric pressure. The pump, the movement of bodies through a resistant medium: Stevin, at the end of the 16th and beginning of the 17th centuries, the engineer and inspector of the land and water equipment of Holland; Galileo, Torricelli (first quarter of 17th century); Pascal (1647-1653); Herique (1650-1663), engineer to the army of Gustavus Adolphus, the builder of bridges and canals. Robert Boyle (seventies of the seventeenth century). Academy del Cimente (1657-1673).

4. Problems of the mechanics of the heavens, the theory of tides. Kepler (1609); Galileo (1609-1616); Gassendi (1647); Wren (sixties of 17th century); Halley (seventies of 17th century); Robert Hooke.

The above specified problems embrace almost the whole sphere of physics.

If we compare this basic series of themes with the physical problems which we found when analysing the technical demands of transport, means of communication, industry and war, it becomes quite clear that these problems of physics were fundamentally determined by these demands.

In fact the group of problems stated in the first paragraph constitute the physical problems relating to raising equipment and transmission mechanism important to the mining industry and the building art.

The second group of problems has fundamental significance for artillery and constitutes the basic physical tasks of ballistics.

The third group of problems is of fundamental importance to the problems of pumping water from mines and of their ventilation, the smelting of ores, the building of canals and locks, intrinsic ballistics and calculating the form of vessels.

The fourth group is of enormous importance to navigation.

All these are fundamentally mechanical problems. This of course does not mean that during this period other aspects of the movement of matter did not occupy attention. During this period optics began to develop and the first observations on static electricity and magnetism were made. (1). Nevertheless both by their nature and by their specific importance these problems have quite a subsidiary significance, and by the extent of their investigation and mathematical development (with the exception of certain laws of geometrical optics, which were of considerable importance in the construction of optical instruments) lagged far behind mechanics.

So far as optics were concerned this science received its main impulse from those technical problems which were of importance first and foremost to marine navigation. (2).

We have compared the main technical and physical problems of the period with the scheme of investigations

1. Investigations into magnetism developed under the direct influence of the study of the deviation of the compass in the world's magnetic field, which had first been met with during the first distant sea expeditions. Gilbert gave much attention to the problems of the earth's magnetism.
2. During this period optics developed through study of the problem of the telescope.

governing physics during the period we are investigating, and we come to the conclusion that the scheme of physics was mainly determined by the economic and technical tasks which the rising bourgeoisie raised to the forefront.

During the period of merchant capital the development of productive forces set science a series of practical tasks and made an imperative demand for their accomplishment.

Official science, the centres of which were the mediæval universities, not only made no attempt to accomplish these tasks, but actively opposed the development of natural sciences.

The universities of the fifteenth to the seventeenth centuries were the scientific centres of feudalism. They were not only the centres of feudal traditions but the active defenders of those traditions.

In 1655 during the struggle of the craft masters with the workers, the Sorbonne actively defended the masters and the craft system, supporting the masters with "proofs from science and holy writ."

The entire system of pedagogy in the mediæval universities constituted a closed system of scholasticism. There was no place for natural science in these universities. In Paris in 1355 it was decided to teach Euclid only on holidays.

The chief "natural-science" manuals were Aristotle's books, from which all the vital content had been removed. Even medicine was taught as a logical science. Nobody was allowed to study medicine unless he had studied logic for three years previously. It is true that when sitting for the medical examination the student had to face a question of a non-logical character, (testimony to his being the child of a lawful marriage) but obviously this one illogical question was hardly sufficient for a knowledge of medicine, and the famous chirurgian Arnold Villeneuve of Montpellier complained that even the professors in the medical faculty were not only unable to cure sufferers from the most ordinary of illnesses, but even unable to apply a leech.

The feudal universities struggled against the new science with a strength equal to that exerted by the dying feudal relationships against the new progressive methods of production.

Whatever was not to be found in Aristotle for them simply did not exist.

When Kircher (the beginning of the 17th century) suggested to a certain provincial Jesuit professor that he should gaze through the telescope at the newly discovered sun-spots, the latter replied: "It is useless, my son. I have read Aristotle through twice and have not found anything about spots on the sun in him. There are no spots on the sun. They arise either from the imperfections of your telescope or from the defects of your own eyes."

When Galileo invented the telescope and discovered the phases of Venus, whilst the merchant companies turned to him for his telescope, which was superior to those made in Holland, the scholastic university philosophers refused to hear about these new facts.

"We must smile, Kepler," Galileo wrote bitterly on August 19th, 1610, "at the great stupidity of men. What are you to say of the first philosophers of the school here, who with the stubbornness of an adder, despite invitations a thousand times repeated, did not wish even to glance either at the planets or at the moon, or even at the telescope itself. Truly the eyes of these men are closed to the light of truth. It is astounding, yet it does not surprise me. This kind of person thinks that philosophy is a kind of book .... that truth has to be sought not in the world, not in nature, but in the collation of texts."

When Descartes resolutely declared himself against Aristotelian physics and against University scholastics he met with savage opposition from Rome and the Sorbonne.

In 1671 the theologians and medicos of the Paris University demanded a governmental decision condemning Descartes' teaching.

In a biting satire Boileau ridiculed these demands of the learned scholastics. This notable document excellently describes the position of affairs in the mediæval universities.

Even in the second half of the 18th century the Jesuit professors in France could not reconcile themselves to Copernicus' theories. In 1760, in the Latin translation of Newton's "Principia," Lesser and Jacquier thought it necessary to add the following note: "In his third book Newton applies the hypothesis of the movement of the earth. The author's assumptions cannot be explained except on the basis of this hypothesis. Thus we are compelled to act in another's name. But we ourselves openly declare that we accept the decisions published by the heads of the church against the movement of the earth."

The universities prepared almost exclusively ecclesiastics and jurists.

The church was the international centre of feudalism and itself was a large feudal proprietor, as not less than one third of the land in Catholic countries belonged to it.

The mediæval universities were a powerful weapon of church hegemony.

Meantime, the technical problems which we have above outlined demanded enormous technical knowledge, and extensive mathematical and physical studies.

The end of the middle-ages, (the middle of the 15th century) is characterised by a higher degree of development of the industry created by the mediæval burghers.

Production now became more perfect, various, and on a larger; a mass scale. Commercial relationships were more developed.

When, remarks Engels, after the dark night of the middle ages science again began to develop at a marvellous speed, industry was responsible.

From the time of the crusades industry developed enormously and had a mass of new achievements to its credit (metallurgy, mining, the war industry, dyeing), which supplied not only fresh material for study, but also new means of experimentation and allowed of the construction of new instruments.

It can be said that systematic experimental science became possible from this time.

Further, the great geographical discoveries, which in the last resort were also determined by industrial interests, supplied an enormous, and previously inaccessible mass of material in the realm of physics (magnetic deviation) astronomy, meteorology, and botany.

Finally, during this period appeared that mighty instrument of the distribution of knowledge: the printing press.

The construction of canals, locks and ships, the construction of shafts and working of mines, their ventilation, the pumping out of water from them, the planning and construction of firearms and fortresses, the problems of ballistics, the production and planning of instruments for navigation, the working out of methods of establishing courses at sea,—all demanded workers of a totally different type from those then being produced by the universities.

By the third quarter of the 16th century, when specifying the minimum of knowledge required by a mine-surveyor, Johann Matesius pointed out that he must have a thorough knowledge of the method of triangulation, must know Euclidian geometry well, must be able to use the compass, necessary in constructing galleries, must be able to calculate the correct direction of the mine, and must understand the construction of pumping and ventilation apparatus.

He pointed out that in order to construct galleries and work the mines theoretically educated engineers were required, since this work was far beyond the powers of an ordinary, uneducated miner.

In view of this it was obviously not possible to learn the profession in the universities of the time. The new science grew up in struggles with the universities, as a non-university science.

The struggle of the university and non-university science serving the needs of the rising bourgeoisie was a reflection in the ideological realm of the class struggle between the bourgeoisie and feudalism.

Science flourished step by step with the development and flourishing of the bourgeoisie. In order to develop its industry the bourgeoisie needed science, which would investigate the qualities of material bodies and the forms of manifestation of the forces of nature.

Hitherto science had been the humble servant of the church, and it was not allowed to pass beyond the bounds established by the church.

The bourgeoisie had need of science and science arose together with the bourgeoisie despite the church. (Engels.)

Thus the bourgeoisie came into conflict with the feudal church.

In addition to the professional schools, (schools for mining engineers and for training artillery officers) the centres of the new science, of the new natural sciences were the scientific societies outside the universities.

In the fifties of the 17th century the famous Florentine Academia del Cimente was founded, taking as its function the study of nature by means of experiment. Among its membership were such scientists as Borelli and Viviani. The Academy was the intellectual heir of Galileo and Torricelli and continued their work. Its motto was *Provare e riprovare.*

In 1645 a circle of natural scientists was formed in London; they gathered weekly to discuss scientific problems and new discoveries, and from this gathering developed the Royal Society in 1661. The Royal Society brought together the leading and most eminent of the scientists in England, and in opposition to the university scholasticism adopted as its motto: 'Nullius in verba.' Active part in the society was taken by Robert Boyle, Bruncker, Brewster, Wren, Halley, and Robert Hooke. One of its most outstanding members was Newton.

We see that the rising bourgeoisie brought natural science into its service, the service of developing productive forces. At that time the most progressive class, it demanded the most progressive science. The English revolution gave a mighty stimulus to the development of productive forces. The necessity arose of not merely empirically resolving isolated problems, but of synthetically surveying and laying a stable theoretical basis for the solution by general methods of all the aggregate of physical problems, set for immediate solution by the development of the new technique.

And since (as we have already demonstrated) the basic complex of problems was that of mechanics* this encyclopaedic survey of the physical problems was equivalent to the creation of a harmonious structure of theoretical mechanics which would supply general methods of resolving the tasks of the mechanics of earth and sky.

The explanation of this work fell to Newton to supply. The very name of his most important work indicates that Newton set himself this particular synthetic task.

In his introduction to the 'Principia' Newton points out that applied mechanics and instruction on simple machinery had been worked out previously and that his task consisted not in "discussing the various *crafts* and in resolving sectional tasks, but in giving instruction on nature, the mathematical bases of physics."

Newton's 'Principia' are expounded in abstract mathematical language and we should seek in them in vain for an exposition by Newton himself of the connection between the problems which he sets and solves with the technical demands out of which they arose.

Just as the geometrical method of exposition was not the method by means of which Newton made his discoveries, but, in his opinion, was to serve as a worthy vestment for the solutions found by other means, so in a work treating of 'Natural philosophy' we cannot expect to find references to the 'low' source of its inspiration.

We shall attempt to show that the 'earthy core' of the 'Principia' consists of just those technical problems which we have analysed above and which fundamentally determined the themes of physical research of the period.

Despite the abstract mathematical character of exposition adopted in the 'Principia' Newton was not only not a learned scholastic divorced from life, but in the full sense of the word was in the centre of the physical and technical problems and interests of his time.

Newton's well-known letter to Francis Aston gives a very clear conception of his wide technical interests. The letter was written in 1669 after he had received his professor-

* Optics also began to develop during this period, but the basic investigations in optics were subordinated to the interests of maritime navigation and to astronomy. It is important to note that Newton came to the study of the spectrum by way of the phenomenon of the chromatic abberration in the telescope.

ship, just as he was finishing the first outline of his theory of gravity.

Newton's young friend, Aston, was about to tour various countries of Europe, and he asked Newton to give him instructions how most rationally to utilise his journey and what especially was worthy of attention and study in the continental countries.

Briefly summarised, Newton's instructions were: diligently to study the mechanism of steering and the methods of navigating ships; attentively to survey all the fortresses he should happen to find, their method of construction, their powers of resistance, their advantages in defence, and in general to acquaint himself with war organisation. To study the natural riches of the country, especially the metals and minerals, and also to acquaint himself with the methods of their production and purification. To study the methods of obtaining metals from ores. To discover whether it was a fact that in Hungary, Slovakia and Bohemia close to the town of Eila or in the Bohemian mountains not far from Silesia there was a river with waters containing gold, also to ascertain whether the methods of obtaining gold from gold-bearing rivers by amalgamating with mercury remained a secret, or whether it was now generally known. In Holland a factory for polishing glass had recently been established; he must go to see it. He must learn how the Dutch protected their vessels from rot during their voyages to India. He must discover whether pendulum clocks were of any use in determining longitude during distant ocean expeditions. The methods of transforming one metal into another, iron into copper for instance, or of any metal into mercury, were especially worth attention and study. In Chemnitz and in Hungary, where there were gold and silver mines, it was said they knew how to transform iron into copper by dissolving the iron in vitriol, then boiling the solution, which on cooling yielded copper. Twenty years previously the acid possessing this noble property had been imported into England. Now it was not possible to obtain it. It was possible that they preferred to exploit it themselves in order to turn iron into copper to sell it.

These last instructions, dealing with the problem of transforming metals, occupies almost half this extensive letter.

That is not surprising. Newton's period was still very rich in alchemic investigations. The alchemists are commonly represented as a kind of magician seeking the philosopher's stone. In reality alchemy was closely bound up with production of necessities and the mystery with which the alchemists were surrounded should not conceal from us the real nature of their researches.

The transformation of metals constituted an important technical problem, since the copper mines of the time were very few, and the war business and the casting of cannon demanded much copper.

The developing commerce made great demands on the means of circulation, and the European gold mines could not cover these demands. Together with the drive to the east in search of gold there was an intensification of the search for means of transforming the common metals into copper and gold.

From his youth Newton had always been interested in metallurgical processes, and he later successfully applied his knowledge and practice in his work at the Mint. He attentively studied the classics of alchemy and made considerable extracts which show his great interest in any and every form of metallurgical process.

During the period immediately preceding his work at the Mint, from 1683 to 1689, he gave much study to Agricola's works on metals, and the transformation of metals was his chief interest.

Newton, Boyle and Locke carried on extensive correspondence on the question of transforming metals and exchanged formulae for the transformation of ore into gold. In 1692 Boyle, who had been one of the directors of the East Indian Company, communicated his formula for transforming metal into gold to Newton.

When Montague invited Newton to work at the Mint he did so not merely out of friendship, but because he highly valued Newton's knowledge of metals and metallurgy.

It is interesting and of importance to note that whilst a rich material has been preserved relating to Newton's purely scientific activities, no material whatever has been preserved relating to his activities in the technical sphere. Not even the materials which would indicate Newton's activities at the Mint have been saved, although it is well known that he did much to perfect the processes of casting and stamping money.

In connection with Newton's bi-centenary Jubilee, Laymann Newell, who especially studied the question of Newton's technical activities in the Mint, requested the director of the Mint, Captain Johnson, for materials touching on Newton's activities in the sphere of the technical processes of casting and stamping. In his reply Captain Johnson said that no materials whatever on this side of Newton's work had been preserved. All that is known is his extensive memorandum to the Chancellor of the Exchequer (1717) on the question of a bi-metallic system and the comparative value of gold and silver in various countries. This memorandum shows that Newton's circle of interests was not restricted to the technical questions of money-production, but extended to the economic problems of currency circulation.

Newton took active part and was an adviser to the commission for the revision of the calendar, and among his papers is a work: "Observations on the revision of the Julian calendar," in which he proposes a radical reform of the calendar.

We cite all these facts in opposition to the tradition which has been built up in literature, which represents Newton as an Olympian standing high above all the 'earthly' technical and economic interests of his time, and soaring only in the empyrean of abstract thought.

It has to be said, as I have already observed, that the 'Principia' certainly afford justification for such a treatment of Newton, which, however, as we see, is absolutely opposed to the reality.

If we compare the circle of interests which was briefly outlined above, we have no difficulty in noting that it embraces almost entirely all that group of problems which arose from the interests of transport, commerce, industry and military affairs during his period, which we summarised.

Now let us turn to an analysis of the contents of Newton's 'Principia' and consider in what inter-relationships they stand with the themes of physical research of the period.

In the definitions and axioms or laws of motion are expounded the theoretical and methodo-logical bases of mechanics.

In the first book is a detailed exposition of the general laws of motion under the influence of central forces. In this way Newton provides a preliminary completion of the work to establish the general principles of mechanics which Galileo had begun.

Newton's laws provide a general method for the resolution of the great majority of mechanical tasks.

The second book, devoted to the problem of the movement of bodies, treats of a number of problems connected with the complex of problems which we have already noted.

The first three sections of the second book are devoted to the problem of the movement of bodies in a resistant medium in relation to various cases of the dependence of resistance upon speed (lineal resistance, resistance proportional to the second degree of speed and resistance proportional to part of the first part of the second degree).

As we have above shown when analysing the physical problems of ballistics, the development of which was connected with the development of heavy artillery, the tasks set and accomplished by Newton are of fundamental significance to extrinsic ballistics.

The fifth section of the second book is devoted to the fundamentals of hydrostatics and the problems of floating bodies. The same section considers the pressure of gases and the compression of gases and liquids under pressure.

When analysing the technical problems set by the construction of vessels, canals, water-pumping and ventilating equipment, we saw that the physical themes of these problems relate to the fundamentals of hydrostatics and aerostatics.

The sixth section deals with the problem of the movement of pendulums against resistance.

The laws governing the swing of mathematical and physical pendulums in a vacuum were found by Huygens in 1673 and applied by him to the construction of pendulum clocks.

We have seen from Newton's letter to Aston of what importance were pendulum clocks in determining longitude. The application of clocks in determining longitude led Huygens to the discovery of centrifugal force and the changes in acceleration of the force of gravity.

When the pendulum clocks brought by Riche from Paris to Caen in 1673 displayed a retarded movement Huygens was able at once to explain the phenomenon by the changes in acceleration of the force of gravity. The importance attached by Huygens himself to clocks is evident from the fact that his chief work is called: 'On pendulum clocks.'

Newton's works continue this course, and just as he passed from the mathematical case of the movement of bodies in a resistant medium with lineal resistance to the study of an actual case of movement, so he passed from the mathematical pendulum to an actual case of a pendulum's movement in a resistant medium.

The seventh section of the second book is devoted to the problem of movements of liquids and the resistance of a thrown body.

In it problems of hydrodynamics are considered, among them the problem of the efflux of liquids and the flow of water through tubes. As was above shown, all these problems are of cardinal importance in the construction and equipment of canals and locks and in planning water-pumping equipment.

In the same section the laws governing the fall of bodies through a resistant medium (water and air) are studied. As we know, these problems are of considerable importance in determining the trajectory of a thrown body and the trajectory of a shot.

The third book of the 'Principia' is devoted to the 'System of the World.' It is devoted to the problems of the movements of planets, the movement of the moon and the anomalies of that movement, the acceleration of the force of gravity and its variations, in connection with the problem of the inequality of movement of chronometers in sea-voyages and the problem of tides.

As we have above indicated, until the invention of the chronometer the movement of the moon was of fundamental importance in determining longitude. Newton returned to this problem more than once (in 1691). The study of the laws of the moon's movement was of fundamental importance in compiling exact tables for determining longitude, and the English 'Council of Longitude' instituted a high reward for work on the moon's movement.

In 1713 Parliament passed a special bill to stimulate investigations in the sphere of determining longitude. Newton was one of the eminent members of the Parliamentary commission.

As we have pointed out in analysing the sixth section, the study of the movement of the pendulum, begun by Huygens, was of great importance to navigation, consequently in the third book Newton studies the problem of the second pendulum, and subjects to analysis the movement of clocks during a number of ocean expeditions: that of Halley to St. Helena in 1677, Varenne's and de Hais's voyage to Martinique and Guadeloupe in 1682, Couple's journey to Lisbon, etc., in 1697, and a voyage to America in 1700.

When analysing the causes of tides Newton subjects the height of flow tides in various ports and river mouths to analysis, and discusses the problem of the height of flows in dependence on the local situation of the port and the forms of the flow.

This rough outline of the contents of the 'Principia' exhibits the complete coincidence of the physical thematics of the period, which arose out of the needs of economics and technique, with the main contents of the 'Principia,' which in the full sense of the word is a survey and systematic resolution of all the main group of physical problems. And as by their character all these problems were problems of mechanics, it is clear that Newton's chief work was a survey of the mechanics of the earth and the heavenly bodies.

THE CLASS STRUGGLE DURING THE ENGLISH REVOLUTION AND NEWTON'S PHILOSOPHIC OUTLOOK.

It would, however, be too greatly simplifying and even vulgarising our object if we began to quote *every problem* which has been studied by one physicist or another, and every economic and technical problem which he solved.

According to the materialistic conception of history, the final determining factor in the progress of history is the creation and recreation of actual life.

But this does not mean that the *economic factor* is the sole determining factor. Marx and Engels severely criticised Barth for narrowing down historical materialism to such a primitive conception.

The economic position is the foundation. But the development of theories and the individual work of a scientist are affected by various superstructures, such as political forms of class war and the results, the reflection of these wars on the minds of the participants—political, juridical, philosophic theories, religious beliefs and their subsequent development into dogmatic systems.

Therefore, when analysing the thematics of physics we took the main, cardinal problems on which the attention of scientists was riveted in that epoch. But, in order to understand how Newton's work proceeded and developed and in order to explain all the features of his physical and philosophic creative powers, the above general analysis of the economic problems of the epoch would not be sufficient. We must analyse more fully Newton's epoch, the class struggles during the English Revolution, and the political, philosophic and religious theories are reflected in the minds of the contemporaries of these struggles.

When Europe emerged from the Middle Ages, the rising town bourgeoisie was its revolutionary class. The position which it occupied in the feudal society had become too restricted for it, and its further free development had become incompatible with the feudal regime.

The great struggle of the European bourgeoisie against feudalism reached its greatest intensity in three important and decisive battles: (1) the Reformation in Germany, with the political rising of Franz Zikkengen and the Great Peasant War which followed it. (2) The Revolution of 1649-1688 in England. (3) The Great French Revolution.

There is, however, a great difference between the French Revolution of 1789 and the English Revolution.

Feudalism in England had been undermined from the times of the Wars of the Roses. The English aristocracy at the beginning of the XVII century was of very recent origin. Out of 90 peers, sitting in Parliament in 1621, 42 had received their peerages from James I, whilst the lineage of the others dated back only to the XVI century.

This explains the close relationship between the higher nobility and the first Stuarts. This feature of the new aristocracy enabled it to compromise more easily with the bourgeoisie.

It was the urban bourgeoisie that began the English Revolution and the middle-class peasant yeomanry brought it to a victorious end.

1689 was the compromise between the rising bourgeoisie and the former great feudal landlords. Far from opposing the development of industry, the English aristocracy of the times of the Henry VII tried, on the contrary, to extract gain from it.

The bourgeoisie was becoming an acknowledged, though a modest section of the ruling classes of England.

In 1648 the bourgeoisie fought, together with the new aristocracy, against the Monarchy, feudal nobility and the dominant Church.

In the Great French Revolution of 1789 the bourgeoisie, in alliance with the people, fought against the Monarchy, the nobility and the dominant Church.

In both Revolutions the bourgeoisie was the class which actually stood at the head of the movement.

The proletariat and those strata of the urban population which did not belong to the bourgeoisie, either did not yet have any interests separate from the bourgeoisie, or did not form an independently developed class or part of a class.

Therefore, wherever they arose against the bourgeoisie as for instance, in 1793-4 in France, they fought only for the realization of the interests of the bourgeoisie, though not in the bourgeois fashion.

The whole of the French terror is nothing but a plebian chastisement of the enemies of the Revolution: absolutism and feudalism. The same may be said of the movement of the Levellers during the English Revolution.

The Revolutions of 1648 and 1789 were not essentially English or French Revolutions. They were, in essence, European revolutions. They not only represented the victory of a single definite class over the old political structure, but they heralded the political structure of a new European society.

"The Bourgeoisie conquered in them. But the victory of the Bourgeoisie then meant the victory of the new social regime, the victory of bourgeois over feudal property rights, the victory of the nation over provincialism, of competition over trade guilds, the division of property over primogeniture, the owner's possession of the land instead of being enslaved to the land, the victory of education over superstition, of family over family name, of industry over heroic indolence, of bourgeois rights over mediæval privileges."

The English Revolution of 1649-1688 was a bourgeois revolution. It gave power to the "profiteers" who had sprung from the capitalists and landowners. The Restoration did not mean at all the re-establishment of the feudal system. On the contrary, in the Restoration the owners of land destroyed the feudal system of land ownership. In essence, Cromwell was doing the work of the rising bourgeoisie. The pauperisation of the population, as the forerunner to the creation of a free proletariat, is particularly marked after a revolution. It is in this change of the ruling class that the true meaning of a revolution is to be found. The new economic system then forming produces a new governing class. Herein lies the main difference between the interpretation of Marx and those of traditional English historians, and particularly those of Hume and Macaulay.

Like a true Tory, Hume views the importance of the Revolution of 1641 and the Restoration, and then the Revolution of 1688, only from the aspect of the destruction and re-establishment of order.

He severely condemns the upheaval caused by the first Revolution and welcomes the Restoration as a means of re-establishing order. He sympathises with the Revolution of 1688 as a constitutional act, although he does not consider that this Revolution brought about the simple restoration of the old freedom. It opens a new constitutional epoch, giving "predominance to the popular element."

To Macaulay the Revolution of 1688 is closely connected with the first Revolution. But the Revolution of 1688 is to him 'the glorious revolution' just because it is a constitutional one.

He wrote his History of 1688 immediately after 1848, and everywhere his fear of proletariat and its possible victory is evident. He relates with proud joy that, when depriving James II of his throne, Parliament observed all the detailed precedents and even sat in the ancient halls in robes prescribed by ritual.

Law and constitution are regarded as non-historical truth unconnected with the dominant class, and thus the way to understanding the actual essence of the Revolution is closed.

Such was the distribution of class forces after the English Revolution. The fundamental philosophical tendencies of the epoch directly preceding the English revolution and following it were: materialism, which originated from Bacon, and was introduced into Newton's epoch by Hobbes, Tolland, Overton and partly by Locke: idealistic sensualism, as presented by Berkeley (H. Moore was closely associated with this): further, a fairly strong trend of moral philosophy and Deism, represented by Shaftesbury and Bolingbroke.

All these philosophic tendencies existed and developed in the complicated conditions of class struggle, the main features of which have been outlined above.

From the time of the Reformation the Church became one of the chief bulwarks of the sovereign power. The Church organisation is a component part of the State system, and the King is the head of the State Church. James I. was fond of saying—"No Bishop, no King."

Every subject of an English King had to belong to the State Church. Anyone not belonging to it was regarded as committing an offence against the State.

The struggle against the absolute power of the King is at the same time a struggle against the centralism and absolutism of the dominant State Church, and therefore the political struggle of the rising bourgeoisie against absolutism and feudalism was carried on under the flag of religious democracy and tolerance.

The collective name of "Puritans" applies to all partisans of the purification and democratisation of the ruling Church. One must distinguish among the Puritans, however, the movement of the more radical Independents from that of the more conservative Presbyterians. These two movements formed the basis of political parties.

The partisans of the Presbyterians were mainly the representatives of well-to-do merchants and the urban bourgeoisie. The Independents drew their supporters from the ranks of the rural and urban democrats.

Thus both the class struggle of the bourgeoisie against absolutism and the struggle of the movements within the ranks of the bourgeoisie and peasantry were waged under the cloak of religion.

The religious tendency of the bourgeoisie was still further strengthened by the development of materialistic teachings in England.

Let us briefly review the main stages of the development of materialism in this epoch and its most important representatives.

Bacon was the originator of materialism. His materialism arose out of the struggle with the mediæval scholastics. He wanted to release humanity from the old traditional prejudices and to create a method for controlling the forces of Nature. In his teachings were hidden the seeds of the manifold development of this doctrine. "Matter smiles with its poetic, sensitive gleam at all humanity" (Marx).

In the hands of Hobbes, materialism became abstract and one-sided. Hobbes did not evolve Bacon's materialism, but only systematised it. Sensuality lost its bright colours and was transformed into the abstract sensuality of a geometrist. All the variety of motion was sacrificed to mechanical movement. Geometry was proclaimed as the dominant science (Marx). The living soul was cut out of materialism, and it became hostile to mankind. This abstract, calculating, formally mathematical materialism could not stimulate to revolutionary action.

That is why the materialistic theory of Hobbes did not interfere with his monarchical views and defence of absolutism. After the victory of the Revolution of 1649 Hobbes went into exile.

But contemporaneously with the materialism of Hobbes there existed another materialistic movement, indissolubly bound up with the true revolutionary movement of the Levellers. At the head of this movement was Richard Overton.

Richard Overton was the loyal companion-in-arms of the leader of the Levellers—John Lilburn, the fiery exponent of revolutionary ideas and brilliant political pamphleteer. In contradistinction from Hobbes, he was a practical materialist and revolutionary.

The fate of this fighter and philosopher is curious. Whilst the name of Hobbes is widely known and to be found in all text-books on philosophy, one cannot find a single word about Overton not only in the most detailed bourgeois primer of philosophy, but even in the most complete biographical encyclopædias.

Richard Overton did not write much. He changed the pen too often for the sword and philosophy for politics. His treatise "Man is mortal in all respects" was published in the first edition in 1643, and the second edition in 1655. It is a strikingly materialistic and atheistic composition. Immediately after its appearance it was condemned and prohibited by the Presbyterian Church.

The manifesto of the Presbyterian Assembly directed against unbelief and false religions calls down all the curses on Richard Overton's head. "The chief representative of the terrible doctrine of materialism," declares the manifesto—"rejecting the immortality of the soul, is Richard Overton, the author of the book on the mortality of man."

We will not go into the details of Overton's doctrine, and his fate—a most interesting page in the history of English materialism, but will only mention one point from the publication mentioned, in which Overton formulated very clearly the basic principles of his materialistic doctrine.

In criticising the contrast of the body as inert matter to the soul as the active, creative principle, Overton writes:

"Form is always the form of matter, and matter is the material for form. Each of them cannot exist by itself alone but only in unity with the other, and only in unity do they form a thing.

"Everything created is created from natural elements (Overton uses the term 'elements' in the sense of the ancient Greeks: water, air, earth). But everything created is material, because that which is not material does not exist."[1]

As distinct from England, materialism on French soil was the theoretical standard of French republicans and terrorists, and formed the basis of the "Declaration of the Rights of Man."

In England the revolutionary materialism of Overton was the teaching of only one extreme group, and the main struggle went on under the cloak of religion.

English materialism as preached by Hobbes proclaimed itself a philosophy most suited for scientists and educated people, as against religion, which was good enough for the uneducated masses, including the bourgeoisie.

Together with Hobbes, cut off from his active revolutionism, materialism went to the defence of the royal authority and absolutism, and encouraged the repression of the people.

Even with Bolingbroke and Shaftesbury the new deistic form of materialism remained an esoteric, aristocratic science.

Therefore the "misanthropic" materialism of Hobbes was hateful to the bourgeoisie, not only because of its religious heresy but because of its aristocratic connections.

Because of this and in opposition to the materialism and deism of the aristocracy, it was the Protestant sects, who produced the cause and the fighters against the Stuarts, who also provided the chief fighting forces of the progressive middle class (Engels).

But still more hateful to the bourgeoisie than the esoteric materialism of Hobbes was the materialism of Overton, a materialism which was the banner of the political struggle against the bourgeoisie, a materialism which approached a militant atheism and which fearlessly opposed the very bases of religion.

Newton was the typical representative of the rising bourgeoisie, and in his philosophy he embodies the characteristic features of his class. We may, with every right, apply to him that characterisation which Engels applied to Locke. He also was a typical son of the class compromise of 1688.

Newton was the son of a small farmer. His position in the University and in society until his appointment as a Warden of the Mint (1699) was a very modest one. By his connections also he belonged to the middle classes. His philosophic relations were nearest to Locke, Samuel, Clarke and Bentley.

In his religious beliefs Newton was a Protestant. He was an ardent supporter of religious democracy and tolerance. We shall see later that the religious beliefs of Newton were a component part of his world-outlook.

In his political views Newton belonged to the Whig Party. During the second revolution Newton was a Member of Parliament for Cambridge from 1689 to 1690. When the conflict over the question of the possibility of taking the oath to "the illegal Ruler"—William of Orange—arose, and matters even developed to the point of disorders in Cambridge, Newton, who as Member of Parliament for the Cambridge University had to take the oath from the University, was in favour of the oath of allegiance and the recognition of William of Orange as King.

In his letter to Doctor Cowell Newton adduced three arguments in favour of taking the oath to William of Orange, which were to remove any doubts as to the possibility of taking the oath by those members of the University who had previously sworn fidelity to the deposed King.

The reasoning and arguments of Newton remind one strongly of the opinions of Macaulay and Hume, which were mentioned above.

This ideological characteristic of Newton, who was the child of his class, explains why those materialistic germs which were hidden in the "Principia" did not grow in Newton into a fully formed structure of mechanical materialism, similar to the Physics of Descartes, but intermingled with his idealistic and theological beliefs, which, in philosophical questions, even subordinated the material elements of Newton's Physics.

The importance of the "Principia" is not confined only to technical matters. Its very name indicates that it forms a system, a conception of the universe. Therefore it would be incorrect to limit the analysis of the contents of the "Principia" to determining its intrinsic connection with the economics and technology of the epoch which served the needs of the rising bourgeoisie.

Modern natural science is indebted for its independence to its freedom from teleology. It recognises only the causative study of Nature.

One of the fighting slogans of the Renaissance was: "True knowledge only through knowledge of Causes" (*vere scire per causas scire*).

Bacon emphasised that the teleological view is the most dangerous of its *idola*. The true relations of things are found in the mechanical causation. "Nature knows only mechanical causation, to the investigation of which all our efforts should be directed."

The mechanistic conception of the universe necessarily leads to a mechanistic conception of causation. Descartes lays down the principle of causation as "an eternal truth."

On English soil mechanistic determinism came to be generally accepted, although it was found interwoven with religious dogma (the sect "Christian necessarians," to which Priestley belonged). This peculiar combination—so characteristic of the English type of thinkers—is found also in Newton.

The universal acceptance of the principle of mechanical causation as the sole and basic principle of the scientific investigation of Nature is due to the mighty development of mechanics. Newton's "Principia" is a grandiose application of this principle to our planetary system. "The old teleology has gone to the Devil," but so far only in the realm of inorganic nature and in the field of terrestrial and celestial mechanics.

The basic idea of the 'Principia' consists of the conception of the movement of the planets as a consequence of the unity of two forces: one directed towards the sun, and the other that of the original impulse. Newton left this original impulse to God.

This unique "division of labour" in the government of the universe between God and causation was characteristic of the English philosophers' interweaving of religious dogma with materialistic principles of mechanical causation.

The acceptance of the modality of movement, and the rejection of moving matter as *causa sui* was bound inevitably to bring Newton to the conception of the original impulse. From this aspect, the conception of divinity in Newton's system is not a casual one, but is organically connected with his views on matter and motion, as well as with his views on space, in the development of which Henry Moore had a great influence on Newton.

It is at this point that the entire weakness of Newton's general philosophic conception of the universe becomes apparent. The principle of pure mechanical causation leads to the understanding of the divine element. "The absurd infinity" of the universal chain of mechanical determinism is closed by the original impulse, and thus the door of teleology is opened.

Thus, the importance of the 'Principia' is not confined to purely physical problems, for it is of great methodological interest.

In the third book of the 'Principia' Newton expounds a "conception of the universe." In the general scholium to the third book (third edition) the indispensability of a divine power is proved as creating, moving and directing elements of the universe.

We shall not go into the question of the authorship of this scholium nor of the role of Cotes and Bentley in the publication of the 'Principia.' A great deal of literature exists on this question, but the letters from Newton quoted below undeniably prove that Newton's theological views were not tacked on to his system and were not forced upon him by Cotes or Bentley.

When Robert Boyle died in 1692 he left a sum yielding £50 per annum, for eight lectures to be read annually in one of the Churches in England, in which proofs of the irrefutability of Christianity were to be given and unbelief repudiated.

Bentley, Chaplain of the Bishop of Worcester, had to read the first series of these lectures. He decided to devote the seventh and eighth to the necessity of the existence of divine providence. He decided to take the proofs for this from consideration of the physical principles of the creation of the world as they are given in Newton's 'Principia.'

When preparing these lectures he met with a series of physical and philosophic difficulties, for the explanation of which he approached the author of the 'Principia.'

In four letters to Bentley Newton replied in detail to Bentley's questions, and these letters provide a valuable source of information on Newton's views on the cosmological problem.

The chief difficulty on which Bentley approached Newton was how to repudiate the materialistic argument brought forward even by Lucretius, that the creation of the world could be explained by purely mechanical principles, if it be assumed that matter possesses its immanently inherent property of gravity and is equally distributed in space.

In his letters Newton pointed out in detail to Bentley how this materialistic argumentation can be overcome.

It is not difficult to see that here it is the question of the theory of the evolution of the universe that is referred to, and on this question Newton is the resolute opponent of a materialistic conception of evolution.

"When I wrote the third book of the 'Principia'" writes Newton to Bentley, "I paid special attention to those principles which could prove to intellectual people the existence of Divine power."

If matter were equally distributed in finite space, then, owing to its power of gravity, it would collect into one large spheric mass. But if matter were distributed in infinite space, then it could, in obedience to the force of gravity, form masses of various magnitudes.

However, in no case is it possible to explain by natural causes how the luminous mass—the sun—is in the centre of the system and actually in the position in which it is situated.

Therefore the only possible explanation is in the acknowledgment of a divine creator of the universe, who wisely distributed the planets in such a manner that they receive the light and warmth necessary to them.

Going further into the question of whether planets, as a consequence of natural causes, can move, Newton pointed out to Bentley that as a consequence of the force of gravity, which is a natural cause, planets can move, but can never achieve periodical rotation on closed orbits, as for this a tangential component is required. Therefore, concludes Newton, in no case is it possible to explain the actual paths of the planets or creation by natural causes, and therefore, on enquiring into the structure of the universe the presence of an all-wise divine element is apparent.

Further, discussing the question of the stability of the solar system, Newton pointed out that such a wonderfully arranged system, in which velocity and mass of bodies are so selected that it retains stable equilibrium, could only be created by a divine mind.

This conception and Newton's appeal to a divine mind as the highest element, creator and prime motive power of the universe, is not in the least accidental, but is the essential consequence of his conception of the principles of mechanics.

Newton's first law of motion attributed to matter the faculty of retaining that state in which it exists.

As Newton considered only the mechanical form of motion his conception of the state of matter is synonymous with the condition of inertia or mechanical transference.

Matter, on which outside forces have no influence, can exist either in a state of inertia or in a state of rectilinear, proportional movement. If a material body is inert, then only an outside force can bring it out of that state.

If, however, a body is in motion, then only an extrinsic force can change that motion.

Thus, movement is not an immanently inherent attribute of a body, but is a modus which matter possibly does not possess.

In this sense Newton's matter is inert in the full meaning of the word. An outside impulse is always necessary to bring it into movement or to alter or end this movement.

Further, as Newton accepts the existence of an absolute, immovable space, to him inertia is possible also as absolute inertia, and thus the existence of absolutely immovable matter, and not merely immovable within the given frame of reference, is physically possible.

It is clear that such a conception of the modality of movement must inevitably lead to the introduction of an extrinsic motive force, and with Newton this rôle is filled by God.

It is very important to note that in principle Newton is not only not opposed to the idea of determining matter by definite attributes, but, contrary to Descartes, declares density and weight to be "immanent qualities of matter."

Thus, denying to movement the character of being an attribute of matter, and accepting it only as a modus, Newton consciously deprives matter of that inalienable property without which the structure, and creation of the world cannot be explained by natural causes.

If we contrast Newton's point of view with that of Descartes, the difference in their beliefs is immediately apparent.

"I say quite openly"—the latter declares—"that in the nature of bodily things I do not recognise any other than that which can be separated in the most distinct manner, can take on form and move, which mathematicians call quantity and make the subject of their demonstrations; that in this matter I consider only its separation, forms and movement and do not accept anything as the truth which does not ensue from these principles as clearly as with the authenticity of mathematical statements. By this means all the phenomena of Nature can be explained. Therefore I hold the view that in Physics other principles from those laid down here are neither necessary nor permissible."

In his physics, Descartes does not admit any supernatural causes. Therefore Marx points out that the mechanistic French materialism was close to Descartes' Physics, in opposition to his metaphysics.

Descartes' Physics could play that rôle only because within the limits of his Physics matter represents a single substance, the only basis of existence and knowledge (Marx).

In the third part of his 'Principia' Descartes also gives a picture of the development of the universe. The difference in position taken up by Descartes consists in his considering in detail the historical genesis of the universal and solar systems in accordance with the principles mentioned above.

It is true that Descarts, also, considers movement only as the modus of matter, but, in contrast to Newton, with him the supreme law is the law of conservation of the quantity of motion. Separate material bodies can acquire and lose movement, but the general quantity of movement in the universe is constant.

In Descartes law of the conservation of the quantity of movement is included the assumption of the indestructibility of movement.

It is true, Descartes understood indestructibility in the purely quantitative sense, and such a mechanical formulation of the law of conservation of movement is not accidental, but arises from the fact that Descartes, like Newton, takes the view that all varieties of movement consist of mechanical transposition. They do not consider the problem of transition from one form of movement to another, and this, as we shall see in the second part of this paper, is for profound reasons.

Engels' great merit is the fact that he considered the process of the movement of matter as eternal transition from one form of material movement to another. This enables him not only to establish one of the basic theses of dialectic materialism, i.e., the inseparability of movement from matter, but also to carry the conception of the law of conservation of energy and quantity of movement to a higher level.

Descartes also introduced God, but his god is necessary to him only to prove that the quantity of movement in the universe remains constant.

He not only does not accept the conception of an outside impulse from God in regard to matter, but, on the contrary, considers that constancy is one of the basic attributes of divinity and therefore in his creations we cannot assume any inconstancy, as by expecting inconstancy in his creations we assume inconstancy in him.

Thus Descartes' reason for introducing a divinity is different from Newton's, but a divinity is also necessary in his conception, as Descartes also does not pursue the view of the self-movement of matter to its logical conclusion.

During the period when Descartes and Newton were working out their conceptions of matter and movement, although somewhat later (the nineties of the XVII. century), we find in John Tolland also a consequential materialistic conception of the correlation of matter and movement.

Criticising the beliefs of Spinosa, Descartes and Newton, Tolland directed his chief attack against the conception of the modality of movement.

"Movement," contended Tolland in his fourth letter to Sirene, "is a most essential attribute of matter, just as inseparable from it as gravity, impenetrability and dimension. It must enter as a component part into its determination."

"This is the only conception"—Tolland quite justly avers, "that provides a rational explanation of the law of the quantitative constancy of movement. It solves all the difficulties regarding motive forces."

The teaching of the self-movement of matter received its full development in the dialetical materialism of Marx, Engels and Lenin.

The entire progress of modern Physics demonstrates the truth of this teaching. In modern Physics, the view of the inseparability of movement from matter is being more and more accepted.

Modern Physics rejects absolute inertia.

As a result of the universally accepted importance of the law of the conservation and transmutation of energy, the conception of the correlation of the forms of movement of matter which was developed by Engels is being more and more confirmed. It is the only conception giving a true understanding of the law of transmutation of energy, as it synthesises the quantitative side of this law with its qualitative side, uniting it organically with the self-movement of matter.

The connection of the law of inertia and the conception of inert matter with Newton's absolute space has been indicated above.

But Newton did not confine himself only to the physical conception of space, but gave also a philosophic-theological conception.

Dialetical materialism considers space as a form of existence of matter. Space and time are the root conditions of the existence of all beings, and therefore space is inseparable from matter. All matter exists in space, but space exists only in matter. Empty space divorced from matter is only a logical or mathematical abstraction, the fruit of the activities of our minds, to which no real thing corresponds.

According to Newton's thesis space can be divorced from matter, and absolute space preserves its absolute properties because it exists independently of matter.

Material bodies are found in space, as in a kind of receptacle. Newton's space is not a form of the existence of matter, but only a receptacle independent of these bodies and existing independently.

Such is the conception of space as laid down in the 'Principia.' Unfortunately, we cannot enter here into a detailed analysis of this conception. We will only note that such a conception is closely connected with the first law of motion.

Having thus determined space as a receptacle, separated from matter, Newton, naturally, asks himself the question what is the essence of this receptacle.

In solving this question Newton concurs with H. Moore, who held the view that space is " the sensorium of God."

In this question also Newton fundamentally differs from Descartes, who developed the conception of space as a physical body.

The unsatisfactory nature of the conception of Descartes lies in the fact that he identified matter with geometric volume.

Whilst Newton separated space from matter, Descartes, by materialising geometrical forms, deprived matter of all properties except extension. This, of course, is also incorrect, but this conception did not lead Descartes in his physics to the same conclusions as Newton.

What is found in space devoid of matter, asks Newton in question 28 in "Optics." How can it be that in Nature everything is consistent and whence arises the harmony of the world? Does it not follow from the phenomena of Nature itself that there is an immaterial, intellectually gifted, omnipresent being for whom space is the sensorium, through which it perceives things and conceives them in their essence?

Thus we see that in this question also Newton decidedly accepts the viewpoint of theological idealism.

Thus the idealistic views of Newton are not accidental, but organically bound up with his conception of the universe.

Whilst in Descartes we find a sharply defined dualism in his physics and metaphysics, in Newton, particularly in his later period, we not only do not find any desire to separate his physical conception from the philosophical, but he even, on the contrary, attempts in his 'Principia' to justify his religio-theological views.

In so far as the Principia' arises in the main from the requirements of the economy and technology of the epoch and studies the laws of the movement of material bodies, the book undoubtedly has elements of healthy materialism.

But the general defects of Newton's philosophic conception outlined above, and his narrow mechanical determinism, not only do not allow Newton to develop these elements, but even on the contrary thrust them into the background to Newton's general religio-theological conception of the universe.

So that in his philosophic as well as in his religious and political views Newton was a child of his class. He ardently opposed materialism and unbelief.

In 1692 Newton, after the death of his mother and the fire which destroyed his manuscripts, was in a state of depression. At that time he wrote to Locke, with whom he corresponded on various theological matters, a sharp letter regarding his philosophic system.

In his letter of the 16th September 1693 he asked Locke to forgive him for this letter and for having thought that Locke's system affects moral principles. Newton particularly asked forgiveness for having considered Locke as a follower of Hobbes.

Here is found the confirmation of Engels' statement that Hobbes' materialism was hateful to the bourgeoisie.

There is no need even to speak of Overton's materialism —he was, after all, almost a Bolshevik.

When Leibnitz, in his letters to the Princess of Wales, accused Newton of materialism because he considered space as the sensorium of divinity, by which he conceives things, which, consequently do not wholly depend on him and are not created by him, Newton fiercely protested against such accusations. Clarke's polemics with Leibnitz had as their object the rehabilitation of Newton from this accusation.

In the realm of physics Newton's researches remain in the main within the bounds of one form of movement—that is, mechanical transposition, and therefore contain no conception of development and transition from one form of movement to another, and in the realm of his views on Nature as a whole the conception of development is entirely absent in Newton.

Newton closes the first period of the new natural science in the field of the inorganic world. It is a period of mastery of the material available. In the realm of mathematics, astronomy and mechanics he achieved great results, particularly as regards the work of Kepler and Galileo, which Newton completed.

But all historical outlook on nature is lacking. As a system it is absent in Newton. Natural science, revolutionary in its origin, comes to a halt in face of conservative nature, which from century to century remains the same as it was created.

Not only is the historical view of nature lacking in Newton, but in his system of mechanics the law of the conservation of energy does not exist. This is all the more incomprehensible, at first sight, in view of the fact that the law of conservation of energy is a simple mathematical consequence of the central forces with which Newton deals.

Further, Newton considers, for instance, cases of oscillation, in explanation of which Huygens, studying the question of the centre of oscillations, gave vague enunciation to the law of the conservation of energy.

It is quite obvious that it was not any lack of mathematical genius or limitation in his physical horizon which prevented Newton from enunciating this law, even in the form of an integral of vital forces.

In order to explain this we must consider the question from the viewpoint of our Marxian conception of the historical process. Such an analysis will enable us to discuss this question in connection with the problem of transition from one form of motion to another, to which the solution was provided by Engels.

ENGELS' CONCEPTION OF ENERGY AND NEWTON'S LACK OF THE LAW OF CONSERVATION OF ENERGY.

In analysing the problems of the inter-relationships of matter and motion in Newton we saw that Tolland took the view that motion was inseparable from matter. Nevertheless, the simple recognition of the inseparability of matter from motion far from resolves the problem of studying the forms of matter's movement.

In nature we observe an endless variety of forms of the movement of matter. If we stop to consider the forms of the movement of matter studied by physics we see that here also are a number of different forms of movement (mechanical, thermal, electro-magnetical).

Mechanics studies that form of motion which consists in the simple passage of bodies through space.

Nevertheless, in addition to this form of motion we have a number of other forms of the movement of matter, in which the mechanical transposition drops to second place by comparison with the new specific forms of motion.

The laws of the movement of electrons, although they are connected with their mechanical transpositions, do not amount to their simple transposition in space.

Consequently, in distinction from the mechanical viewpoint, which regards the main task of natural science as the reduction of all the complex aggregation of the movements of matter to one form of mechanical transposition, dialectical materialism regards the main task of natural science as the study of the forms of movement of matter in their inter-connections, inter-relationships and development.

Dialectical materialism understands movement to be change in general. Mechanical transposition is only one, partial form of movement.

In real matter, in nature we never meet with absolutely isolated pure forms of movement. Every real form of movement, including, of course, mechanical transposition, is always bound up with the transition of one form of movement into another.

Hitherto physics has remained within the bounds of studying one form of movement, the mechanical form, and as we have seen this constitutes the peculiarity of Newton's physics; the problem of inter-relationships between this form and other forms of movement could not truly be set. And when such a problem was set there was always a tendency to hypostatise just this most simple and most fully studied form of movement and to put it forth as the sole and universal aspect of motion.

Descartes and Huygens took up this position, and Newton essentially associated himself with it.

In the introduction to the 'Principia' Newton directs attention to the circumstance that "it would be desirable to deduce from the elements of mechanics the remaining phenomena of nature." (Newton deduced the motion of the planets from these laws in the third book.) "A great deal forces me to assume," he continues, "that all these phenomena (of nature) are conditioned by forces, by which the particles of bodies, in consequence of causes so far unknown, are either attracted one towards another and accumulate in a true figure, or else are mutually repelled and separate one from another."

With the development of large-scale industry the study of the new forms of movement of matter and their exploitation for the needs of production come to the forefront.

The steam engine gave a mighty impulse to the development of the study of the new, thermal form of movement. The study of the history of the development of the steam engine is of importance to us in two regards.

First and foremost we study the problem why it was that the development of industrial capitalism and not that of merchant capital raised the problem of the steam engine. This will explain why the steam engine became the central object of investigation not in Newton's time but in the period immediately following, although the invention of the first steam engine dates from Newton's period. (Ramsay's patent in 1630).

Thus we see that the connection between the development of thermo-dynamics and the steam-engine is the same as that between the technical problems of Newton's period and his mechanics.

But the development of the steam engine is of interest to us in another direction. In distinction from mechanical machines (the block, the windlass, the lever) in which one aspect of mechanical movement is transformed into another aspect of the same mechanical transposition, by its very essence the steam engine is based on the transformation of one form of movement (thermal) into another form (mechanical).

Thus, together with the development of the steam engine we get inevitably also the problem of the transition of one form of movement into another, which we do not find in Newton and which is closely bound up with the problem of energy and its transformation.

We first turn to a study of the chief stages of development of the steam engine in connection with the development of productive forces.

Marx noted that the mediæval commerce of the first trading cities was of an intermediary character. It was founded on the barbarism of the producing peoples, for whom the trading cities and merchants played the rôle of middlemen.

So long as merchant capital played the rôle of middleman in the exchange of produce of undeveloped countries merchant profit was not merely the result of cheating and trickery, but directly originated from them. Later merchant capital utilised the difference in price between the prices of production of various countries. In addition, as Adam Smith emphasises, during the first stage of its development merchant capital is chiefly a contractor and supplies the needs of the feudal landowner or the eastern despot, concentrating the main mass of surplus product in its own hands and being comparatively little interested in the prices of commodities.

This explains the enormous profits of the mediæval trade. The Portuguese expedition of 1521 purchased cloves for two or three ducats and sold them in Europe at 336 ducats. The total cost of the expedition amounted to 22,000 ducats, the receipts were 150,000 ducats, the profits 130,000, i.e., about 600 per cent.

In the beginning of the 17th century the Dutch purchased cloves at 180 guldens for 625 pounds, and sold them in the Netherlands for 1,200 guldens.

The greatest percentage of profit came from those countries which were completely subject to Europeans. But even in the trade with China, which had not lost its independence, the profits reached 75 to 100 per cent.

When the overwhelming hegemony belongs everywhere to merchant capital it constitutes a system of despoliation.

The high rates of profit were maintained in the 17th and the beginning of the 18th centuries. This is to be explained by the circumstance that the extensive trade of the late middle ages and the beginning of the new times was mainly monopolistic commerce. The British East Indian Company was closely connected with the State government. Cromwell's navigation act strengthened the monopoly of British trade. The gradual decline of Holland as a naval power dates from that time and a sound basis is laid to England's maritime hegemony.

Thus, so long as the dominant form of capital was merchant capital, chief attention was directed not so much to the improvement of the actual process of exchange, but to the consolidation of the monopolistic position and to domination in the colonies.

Developing industrial capitalism at once turned its attention to the process of production. The free competition within the country which the British bourgeoisie achieved in 1688 forced an immediate consideration of the problem of costs of production.

As Marx observed, large-scale industry universalised competition and made protective tariffs simply a palliative.

It is necessary not only to produce commodities of good quality and in sufficient quantities, but to produce them as cheaply as possible.

The process of cheapening production of commodities was directed along two lines: the continually increasing exploitation of labour power (the production of absolute surplus value) and the improvement of the production process itself (relative surplus value). The invention of machines not only did not reduce the labour day but on the contrary, being a mighty weapon for the increase of the productivity of labour, as an instrument of capital, it simultaneously became the means of an immeasurable extension of the working day.

We shall trace this process in the steam engine. But before turning to an analysis of the history of the development of the steam engine we must elucidate what we mean by machine, since on this question there exists a radical difference between the point of view of Marxism and that of other investigators.

Meantime, in order to elucidate the essence of the industrial revolution which raised the steam engine to one of the foremost places, it is necessary to have a clear understanding of the rôle played by the steam engine in the industrial revolution.

There is a very widely held view that the steam engine created the industrial revolution. Such an opinion is erroneous. Manufacture develops out of handicrafts by two roads. On the one hand it arises from a combination of heterogeneous independent handicrafts, which lose their independence, on the other hand it arises from the co-operation of craftsmen in the same craft, disintegrating the particular process into its component parts and passing to a division of labour within manufacture.

The starting point in manufacture is labour power.

The starting point in large-scale industry is the means of labour. Of course in manufacture also the problem of the motive power is an important one, but the revolutionisation of all processes of production which was prepared by a detailed division of labour within the bounds of manufacture came not from the motive power but from the executive mechanism.

Every machine consists of three basic parts: the motive power, the transmission mechanism and the executing instrument. The essence of an historical view of the definition of a machine consists in the fact that in various periods a machine has various purposes.

The definition of a machine given by Vitruvius was preserved down to the industrial revolution. For him a machine was a wooden instrument of the greatest service in the lifting and transport of weights.

Consequently the basic contrivances serving these ends: a sloping plane, the windlass, the block, the lever, received the name of simple machines.

When analysing in the 'Principia' the nature of the applied mechanics worked out by the ancients Newton attributes to them the teachings of five simple machines: the lever, the wheel, the block, the windlass, the wedge.

Hence arises the opinion found in English literature that an instrument is a simple machine and a machine a complex instrument.

But it is not entirely a question of simplicity and complexity. The essence of the matter consists in the fact that the introduction of executing mechanism, the function of which consists in seizing and expediently changing the object to be subjected to labour, brings about a revolution in the very process of production.

The two other parts of the machine exist in order to bring the executing mechanism into motion.

Thus it is clear what a gulf divides the machines known to Vitruvius and which accomplish only the mechanical transposition of the finished products, and the machine of large-scale industry, the function of which consists in the complete transformation of the original material of the product.

The fruitful nature of Marx's definition is especially clear if we compare it with the definitions of a machine found in literature.

In his " Theoretical Kinematics " Releau defines a machine as the combination of bodies capable of resisting opposition, and which are built so that by means of their mechanical power the powers of nature are compelled, given certain movements, to bring about an activity.

This definition is equally applicable to Vitruvius' machine and to the steam engine. Although when applying it to the steam engine we meet with difficulties.

The same defect distinguishes the definition of a machine given by Sombart. Sombart calls the machine a means or a complex of means of labour, tended by a man, the purpose of which is the mechanical rationalisation of labour. The machine as a means of labour is distinguished from the instrument of labour by the circumstance that it is tended by a man, whereas the machine as an instrument attends a man.

The unsatisfactory nature of this definition consists in its making the basis of the difference between an instrument and a machine the circumstance that the one serves a man and the other is served by man. This definition, based at first sight on a social economic symptom, not only gives no idea of the difference between the period in which the simple instrument predominates and the period in which the machine method of production predominates, but creates quite an absurd idea that the essence of the machine consists in its being served by man.

Thus an imperfect steam engine demanding the continual service of a man (in Newcomen's first machines a boy had continually to open and close a tap) will be a machine, while a complex automaton producing bottles or electric lamps will be an instrument, since it essentially hardly requires attending.

Marx's definition of a machine directs attention to the circumstance that it causes a revolution in the very process of production.

Motive power is a necessary and very important component part of the machinery of industrial capitalism, but it does not determine its fundamental character. When John Wyate invented his first spinning machine he did not even mention how it was set in motion. "A machine in order to spin without the aid of fingers" was his programme.

Not the development of the motor and the invention of the steam engine created the industrial revolution of the 18th century, but on the contrary the steam engine gained such enormous importance just because the division of labour developing in manufacture and the increasing productivity made it possible and necessary to invent an accomplishing instrument, and the steam engine, which had been born in the mining industry, found a field awaiting its application as a motor.

Arkwright's spinning jenny was at first set in motion by means of water. Meantime the employment of water power as the predominant form of motive power was accompanied with great difficulties.

It was impossible to raise it to a productive level, it was impossible to overcome its defects, sometimes it was exhausted, and it always retained a purely local significance.

Only with the invention of Watt's machine did the machine textile industry, already developed sufficiently, receive the motor without which it could not manage at the stage of development it had reached.

Thus the machine textile industry is not in the least a consequence of the invention of the steam engine.

The steam engine was brought to light in connection with mining. As early as 1630 a patent was granted in England to Ramsay for "raising water with the aid of fire during deep mining works."

In 1711 a "Society for raising water with the aid of fire" was formed for exploiting Newcomen's machine in England.

The greatest service, writes Carnot, in his work "On the moving power of fire," rendered by England's thermal (steam) engine is undoubtedly the revival of the activities of coal mines, which threaten to choke owing to the continually growing difficulties of pumping the water and raising the coal.

The steam engine gradually becomes an important factor in production. Then attention is at once directed to what can be done to make the machine more economical by reducing the expenditure of steam, and consequently the expenditure of water and fuel.

Even before Watt's work Smeaton was occupied with investigating the expenditure of steam in various steam engines, founding a special laboratory in order to do so in 1769. He found that the expenditure of steam varies according to the machine from 176 to 76 kg. per horse-power hour. Savory succeeded in building a machine of the Newcomen type with an expenditure of steam of 60 kg. per horse-power hour.

By 1767 fifty-seven steam engines with a total power of 1,200 horse power were at work around Newcastle alone.

It is obvious that the problem of economy was one of the most fundamental problems confronting Watt.

Watt's patent, taken out in 1769, begins thus: "My method of diminishing the consumption of steam in fire machines, and thus the expenditure of burning material, consists in the following basic propositions."

The agreement which Watt and Bolton concluded with the owner of coal-mines consisted in their receiving in monetary form one-third of the sum received by the saving of expenditure on fuel.

Under this condition from one mine alone they received over two thousand pounds in one year.

The chief inventions of the textile industry were made during the period 1735 to 1780, and thus a potential demand for motors already existed.

In his patent taken out in 1784 Watt described the steam engine as a universal motor of large industry.

The problem of technical rationalization of the steam engine became a central one. The realization of this task in practice made necessary a detailed study of the physical processes carried out in the machine.

In distinction from Newcomen, in the laboratory of Glasgow University Watt studied the thermo-dynamic qualities of steam in detail, and thus laid the basis for thermodynamics as a section of physics.

He carried out a number of experiments on the temperature of boiling water under various pressures in connection with the changes in elasticity of steam. Then he investigated the latent temperature of steam formation and developed and checked Black's theory.

Thus the chief problems of thermodynamics, the teaching on the latent temperature of steam formation, the dependence of the boiling point on pressure and the height of the latent temperature of steam formation began to be scientifically worked out by Watt.

It was this detailed study of the physical processes in the steam machine that enabled Watt to go further than Smeaton, who, although he set himself the task of the laboratory investigation of the steam machine, could not go beyond the purely empirical, superficial improvement of Newcomen's machine, as he was not acquainted with the physical qualities of water vapours.

Thermodynamics not only received an impulse to its development from the steam machine, but, in fact, developed by the study of that machine.

The necessity arose not only of studying the separate physical processes in the steam machine, but the general theory of steam machines, the general theory of the coefficient of profitable activity of the steam machines. This work was carried out by Carnot.

The general theory of the steam machine and the theory of the coefficient of profitable activity led Carnot to the necessity of investigating the general thermic processes, to the discovery of the second element of thermodynamics.

The study of steam machines, said Carnot in his work "On the Motive Power of Fire" (1824) is extraordinarily interesting, as their importance is very considerable and their employment increases with every day. Clearly they will cause a great revolution in the civilized world.

Carnot remarks that, despite various kinds of improvements, the theory of the steam machine had made but little progress.

Carnot formulated his task of discovering the theory of the steam machine in such a way that the practical tasks set by him in order to discover the general theory of the coefficient of useful service were quite clear.

The question is frequently asked, he says, whether the motive power of heat is limited or unlimited; by motive power we mean the useful service which a motor can provide.

Is there any limit to the possible improvements, a limit which the nature of things renders insurpassable by any means whatever? or, on the contrary, are unlimited improvements possible?

Machines that do not derive their motion from heat, but have the motive power of man, animals, the fall of water, the current of air, can be studied, Carnot observes, by means of theoretical mechanics.

Here, all possibilities are foreseen, all possible movements are reduced to general principles (this was made possible owing to Newton's work on mechanics), are firmly established and applicable in all circumstances.

No such theories exist in the case of thermal machines. It will be impossible to establish them, Carnot declared, until the laws of physics are sufficiently extended and sufficiently generalized to make it possible to see in advance the results of a definite reaction of heat on any particular body.

Here the connection between technology and science, between the investigation of the general laws of physics and the technical problems raised by economic development is established with extraordinary clarity.

But the history of the steam machine is important to us in another connection also.

The historical succession in the study of various forms of physical motion of matter is: mechanics, heat, electricity.

We have seen that the development of industrial capitalism faced technology with the demand for the creation of an universal motor.

This demand was preliminarily supplied in the steam machine, which had no competitors until the invention of the electric motor.

The problem of the theory of the coefficient of useful service of steam machines led to the development of thermodynamics, i.e., to the study of the thermal form of movement.

This consequently explains the historical succession in the study of forms of movement; following on mechanics we get the development of the study of thermal forms of motion: thermodynamics.

We now pass to a consideration of the importance of the steam engine from the aspect of the transformation of one form of motion into another.

Whilst Newton never considered even the problem of the law of the conservation and transformation of energy, Carnot was compelled, although truly in an indirect form, to consider it.

This was just because Carnot engaged in the study of the steam engine from the aspect of the transformation of thermal into mechanical energy.

The category of energy as one of the basic categories of physics appears when the problem of the inter-relationships between various forms of motion comes to the forefront. And the more the richness of the forms of motion becomes the subject of study in physics, the greater the importance acquired by the energy category.

Thus the study of physical forms of motion of matter and their historical development must provide the key to an understanding of the origin, importance, and mutual connections of the categories of physics.

The historical study of the forms of motion must be carried on from two aspects. We must study the historical succession of the forms of motion as they appear in the development of physical science in human society. We have already shown the connection between the mechanical and the thermal form of motion from the aspect of their historical genesis in human society. The study of these forms proceeds in the succession that they are raised by human practice.

The second aspect is the study of the "natural science of the development of matter." The process of studying the development of inorganic matter in the microcosmos and the macrocosmos must provide the key to the understanding of the connection and mutual transitions of one form of motion of inorganic matter into another, and must lay a sound basis for the natural classification of the forms of motion of matter. This principle must lie at the basis of Marxist classification.

Every science analyses a separate form of motion or a number of forms of motion connected with one another and passing into one another.

The classification of sciences is nothing more than a hierarchy of the forms of motion of matter in accordance with their essential order, in other words, in accordance with their natural development and the transition of one form of motion into another, as accomplished in nature.

Thus this principle of Marxist classification of science sets at the basis of classification the great idea of development and the transition of one form of matter in motion into another form. (Engels.)

Herein consists Engel's notable conception of the interconnection and the hierarchy of the forms of movements of matter.

The conception of energy is indissolubly bound up with the transformation of one form into another form of motion, with the problem of the measurement of this transformation. Modern physics emphasises the quantitative aspect of this transformation and postulates the constancy of energy through all its transformations.

We recall, as was shown in the previous chapter, that the quantitative constancy and the quantities of motion were announced by Descartes. The new element that was introduced into physics by the work of Mayer and Helmholtz consisted in the discovery of the transformation of the forms of motion and also with the constancy of energy during these transformations.

It was this, and not the simple postulation of constancy that was the new element.

Owing to this discovery the various isolated forces of physics (heat, electricity, mechanical energy) which hitherto could be compared with the invariable forms of biology, were transformed into forms of motion inter-connected and passing into one another according to definite laws.

Like astronomy, physics came to the inevitable conclusion that the last result was the eternal circle of moving matter. That is why Newton's period, which worked only with one form of movement—the mechanical—and put in the forefront not the conversion of one form into another, but only the transformation and modification of one and the same form of motion—mechanical transposition—(we recall the definition of a machine given by Vitruvius and Carnot's observations) did not consider and could not consider the problems of energy.

As soon as the thermal form of motion appeared on the scene (and it appeared on the scene as indissolubly bound up with the problem of its conversion into mechanical motion) the problem of energy came to the forefront. The very setting of the problem of the steam machine (to raise water by means of fire) clearly point to its connection with the problem of the conversion of one form of motion into another. It is significant that Carnot's classic work has the title: "On the Motive Force of Fire."

This treatment of the law of the conservation and conversion of energy given by Engels, raises to the forefront the qualitative aspect of the law of conservation of energy, in contradistinction to the treatment which predominates in modern physics and which reduces this law to a purely quantitative law—the quantity of energy during its transformations. The law of the conservation of energy, the teaching of the indestructibility of motion has to be understood not only in a quantitative but also in a qualitative sense. It contains not only a postulation of the indestructibility and the increatability of energy, which is one of the basic prerequisites of the materialistic conception of nature, but a dialectical treatment of the problem of the movement of matter. From the aspect of dialectical materialism the indestructibility of motion consists not only in the circumstances that matter moves within the limits of one form of motion, but also in the circumstance that matter itself is capable of all the endless variety of forms of motion in their spontaneous transitions one into another, in their self-movement and development.

We see that only, the conception of Marx, Engels and Lenin provides a key to an understanding of the historical succession of development and the investigation of the forms of motion of matter.

Newton did not see and did not solve the problem of the conservation of energy, but not because his genius was insufficiently great. Great men, no matter how notable their genius, in all spheres formulate and resolve those tasks which have been raised for accomplishment by the historical development of productive forces and production relationships.

THE MACHINE-WRECKERS OF NEWTON'S EPOCH AND THE PRESENT-DAY WRECKERS OF PRODUCTIVE FORCES.

We have come to the end of our analysis of the 'Principia.' We have shown how its physical content arose out of the tasks of the epoch, which were raised for accomplishment by the class entering into power.

The historically inevitable transition from feudalism to merchant capital and manufacture, and from manufacture to industrial capitalism, stimulated the development of forces to an unprecedented extent, and this in its turn gave a powerful impetus to the development of scientific research in all spheres of human knowledge.

Newton happened to live in this very epoch, when new forms of social relations, like new forms of production, were being created.

In his mechanics he was able to solve that complex of physico-technical problems which the rising bourgeoisie had set for decision. But he remained impotent before nature as a whole. Newton knew the mechanical transposition of bodies, but he even rejected the conception that nature finds itself in process of unceasing development. Still less can we hope to find in him any view of society as a developing entity, although it was specifically the transitional character of the epoch which gave rise to his basic work.

Has the movement of the historical process ceased since Newton's time? Of course not, for nothing can check the forward movement of history.

After Newton, Kant and Laplace were the first to make a breach in the conception of nature as eternal and unchanging from century to century.

They were to show, albeit in a far from complete form, that the solar system is the product of historic development.

In their works the conception of development, which was subsequently to become the basic and guiding principle of all teaching on nature, entered into natural science for the first time.

The solar system was not created by God, the movement of the planets is not the result of a divine impulse. It not only preserves its condition solely as a consequence of natural causes, but only came into existence through their influence. God is not only unnecessary in a system existing on the basis of the laws of mechanics, but he is unnecessary even as an explanation of its origin.

"I have not found it necessary to include any hypothesis of deity in my system," so Laplace is said to have answered Napoleon's questions as to the reason for the omission of all reference to the rôle of God in his "System of the World."

The progressive development of productive forces gave rise to progressive science. That change from home handicraft industry to manufacture and from manufacture to large-scale machine industry which had only begun during Newton's period, was greatly accelerated during the following century. It was completed during the monopolistic imperialist phase of capitalism, which in turn is introductory to new, socialist forms of development.

As one phase of the capitalist method of production is replaced by another, so the very views of the governing class in capitalist society on technology and science change. Accordingly, on coming to power the bourgeoisie struggles mercilessly against the old guild and handicraft form of production. With an iron hand it introduces large-scale machine industry, shattering in its course the resistance of the worn-out feudal class and the still elemental protest of the new-born proletariat.

For the bourgeoisie science and technology are powerful weapons of struggle, and it is interested in the development and perfection of these weapons.

The glorifier of industrial capitalism (Ure) describes the struggle of the bourgeoisie for new methods of production in the following terms:—

"The horde of discontented, who considered themselves invincible behind the old methods of division of labour, saw that they were out-manœuvred by a flank-attack, and their defensive means were destroyed by modern mechanical technique. They were forced to surrender to the mercy and wrath of the victor."

Examining further the significance of the invention of the spinning machine, he said: "This machine was destined to restore order between the industrial classes. This invention confirms the doctrine which we have already developed, that capital is continually forcing science to serve it and is forcing the rebellious hands of labour into submission."

The bourgeoisie in power talked with the lips of Ure, as it built new methods of production with the flesh and blood of the "rebellious hands of labour."

On coming to power the bourgeoisie revolutionised the whole method of production. It rent the old feudal bonds to shreds, and shattered the archaic forms of social relations which fettered the further development of productive forces. During that period it was revolutionary, because it brought with it new and improved methods of production.

Over a period of a century it changed the face of the earth and brought into existence new, powerful, productive forces.

New, hitherto unexplored forms of movement of matter were discovered.

The gigantic development of technique tremendously stimulated the development of science, and the turbulently developing science in turn permeated the new technique.

And on the basis of this unprecedented flourishing of productive forces, on the basis of the tremendous growth of material culture, occurred an unprecedented impoverishment of the mass of the people, and a terrible growth of unemployment.

It is not strange that these contradictions in the predominant capitalist methods of production should have attracted the attention, not only of State officials and of capitalists, but also of the scientists.

In Newton's epoch the bourgeoisie called for new methods of production. In his memorandum on the reform of the Royal Society, Newton called on the State authorities to support science, which did so much in the study of nature and the creation of new productive forces.

To-day a very different situation obtains.

During the past year "Nature" has published a number of leading articles dealing with the questions we are considering. In these articles problems which are now agitating the whole world receive consideration. Of these articles, we will refer to two which more clearly express the point of view of English natural scientists. One is entitled "Unemployment and Hope," the other "Science and Society."

This is how these articles describe the tasks of industry, its aims and lines of development. Discussing the question of unemployment, in an analysis of capitalist society "Nature" thus defines the rôle of machines.

"There is, indeed, in the present situation much to excuse a passing reflection that perhaps, after all, the people of Erewhon were wiser than ourselves in destroying their machines, lest, as Marx predicted, the machines reversed the original relation and the workmen became the tool and appendage of a lifeless mechanism."

Modern science and technique creates machines remarkable for their accuracy and productivity, of extraordinarily complicated and delicate organisation. And, it appears that the machine-wreckers of Newton's period were wiser than we, who create machines of unprecedented complexity and power.

In the statement referred to not only is there a distortion of the ideas of Marx, but also incorrect light is thrown on the movement of the machine-wreckers.

Let us first re-establish the true historic circumstances and actual causes which provoked the workers into wrecking the machines.

The struggle of the workers against the machine is only the reflection of the struggle between wage labourers and the capitalists. Not against the machines as such did the working class of that period struggle, but against the position to which the developing capitalist order was relegating them in the new society.

During the 17th century almost all Europe seethed with the indignation of the workers against the carding machines. The first wind-power saw-mill was destroyed in London at the end of the " 70's " of the 17th century.

The first decade of the 19th century was notable for the mass movement of the Luddites against the power loom. As industrial capitalism developed it transformed labour power into a commodity. Forced out of industry by machinery, the worker could not find a purchaser for his labour, and was comparable to paper money which has gone out of currency. The growing working-class, still without class consciousness, directed its hatred towards the superficial forms of expression of capitalist relationships—the machines.

But this protest, reactionary in its form, was the expression of a revolutionary protest against the system of wage labour and private ownership of the means of production.

The worker actually became an accessory to the machine not because machines had been invented, but because these machines served the interest of the class owning the means of production.

The call to machine-wrecking will always be a reactionary slogan, and the wisdom of the inhabitants of Erewhon consisted not in their destruction of the machines, but in their protest against the slavery of wage labour.

" The comfort and welfare of the few," continues the leading article, " on this view, may, however, be too dearly purchased when we consider the lot of the displaced workers, and, perhaps, still more the repression of individuality and the retarded development which, as Marx predicted, have often accompanied mass production."

Thus, in the opinion of " Nature," improvement in the means of production inevitably leads to the crushing of individuality and the suffering of the masses of the people.

Here it is permissible to ask : Why was it that during Newton's time, when there was a great development in the means of production, scientific circles not only did not call for a retardation of this development, but, on the contrary, in every way encouraged every new discovery and invention ; and the organ of the foremost natural scientists of Newton's epoch, " Philosophical Transactions," was full of descriptions of these new inventions?

Before answering this question we will see what methods this journal of British naturalists proposes for solving the crisis of production and unemployment, which, according to its views are the results of too great a development of productive forces.

These methods are outlined in the leading article, "Unemployment and Hope." We quote the corresponding section in extenso:—

"The aims of industry are, or should be, chiefly two: (1) to furnish a field for . . . growth of character; and (2) to produce commodities to satisfy man's varied wants, mostly of a material kind, though of course there are large exceptions outside the material category, and the term 'material' is here used in no derogatory sense. Attention has hitherto been directed mainly to (2), and the primary aim of industry has been ignored. Such one-sided view of industry coupled with a too narrow use of the much-abused word 'evolution' . . . has led to over-concentration on quantity and mass production and a ridiculous neglect of the human element, and there can be no doubt that had a little thought been given to the first aim then the second would have been much more completely and satisfactorily attained; also unemployment would not have been heard of . . .

"The prevailing idea . . . appears to be that industry is evolving and must evolve towards one fixed type, for example, that of large-scale production . . . The best form or type of industry . . . may consist of many different and constantly changing forms, distinguished above all things by adaptability and elasticity—a living organism.

"Elasticity further means the possibility of reviving, under new and improved forms to meet modern conditions, two at least of the older types of industry which are supposed to have been superseded or rendered obsolete by modern large-scale production, namely: (1) small cottage industries or handicrafts; . . . (2) a combination of manufacturing with agricultural or garden industry. . . . Industry still has its roots firmly and deeply rooted in the past, and foolishly to tear up a great part of those roots as old and useless is the surest way to weaken the industrial tree. Perchance the source of the unemployment curse is to be found here.

"The restitution of these two principles of an older industrial order, so essentially and characteristically English, under improved forms made possible by modern scientific achievement, including notably electrical power distribution, would furnish, in the first place, a new and almost infinite field for human employment of all kinds, absorbing all or most of the present unemployed . . . By unemployment we mean chiefly the unemployed in Great Britain, but it would be vastly better to extend our consideration to cover unemployment throughout the whole world . . .

" The application of these two principles to unemployment is, of course, only one part of their scope, for they have a far wider range even than this, especially in counteracting one of the greatest evils of modern industry, namely, extreme specialism, monotonous work, and lack of scope for developing skill, with all that that implies . . .

" It is probable that, under the more bracing atmosphere of varied work and interest and skill thus envisaged, the inventive faculties of mankind would be greatly stimulated, and a much-needed spur be given to originality."

Thus, according to " Nature," the remedy for healing the wounds of capitalist society, the methods which are to remove all the contradictions of a system based on wage labour and individual ownership of the means of production, are a return to those forms of industry which directly preceded the epoch of industrial capitalism.

We have demonstrated above that it is from these very forms that the forward movement of the period of Newton began; and although, by comparison with feudal methods of production, manufacture and small handicraft industry were a step forward, at the present moment the slogan—" Back to small handicraft industry "—is deeply reactionary.

The fetishism of the commodity system, laid bare by Marx's genius, lies in the fact that the relationships of material things. created by human society, are isolated from human relationships and are looked upon as the essence of the things themselves.

The solution and exposure of such fetishism consists in the fact that it is not things which of themselves create relations, but that the relationships between things created in the process of social production simply express the specific social relationships of human beings, which in the latter's view take on the fantastic form of relationships between things.

The views cited above also are a special manifestation of fetishism. Machinery, the means of production, the organisation of production in large-scale machine production are considered in isolation, irrespective of the social relationships of that particular economic system in which the specific method of production exists and by which they were created.

The improvement of the instruments of labour brings misfortune to the great mass of the population, we are told. The machine transforms a worker into its mere accessory. It kills individuality. Let us return to the good old times.

No, we reply. It is not the improvement in the means of production that causes the impoverishment and unprecedented sufferings of the masses. It is not the machines which transform the worker into a blind tool of mechanism, but those social relationships which so exploit machinery, that the worker merely becomes an accessory.

The way out lies not in the return to the old out-worn method of production, but in the alteration of the whole system of social relationships just as radically as the transition from feudal and handicraft methods of production to industrial capitalism was effected in the past.

Private property passes through stages of development; feudalism, merchant capital and manufacture, industrial capitalism. At every stage of development in the process of production, men, independently of their will, enter into specific production relationships which correspond to the equivalent stage in the development of productive forces. At a certain stage of their development productive forces come into antagonism with the existing production relationships, or, juridically expressed, with the property relationships, within which they developed. From being forms of development the latter become fetters.

The further development of productive forces is only possible through a radical reconstruction of all production relations.

The transition from one form of production to another is characterised first and foremost by such a reconstruction.

At every new stage the change in social relationships evokes a further turbulent growth of productive forces.

On the contrary, a crisis in the growth of productive forces indicates that their further development within the framework of the given social system is impossible.

And that suggested solution which we cited above, the substance of which consists in the bridling of productive forces by a return to the old forms of production, is only an expression of the contradiction between the productive forces of capitalist society and the production relationships based on private ownership of the means of production.

Science develops out of production, and those social forms which become fetters upon productive forces likewise become fetters upon science.

Genuine methods for the transformation of society cannot be found, through brilliant inspiration or guesswork, and not through a return to " the good old times " which in the distant historical perspective appear to be a peaceful idyll, but which in reality represented bitter class struggle and the crushing of one class by another.

Thus it has always been, and so it was in that epoch when

Newton lived and created, in the epoch to whose productive forms we are invited to return.

We have seen that the outworn forms of social relationships of that epoch, speaking through the lips of their University representatives, also recommended the suppression of science, which was shattering the stagnant forms of feudal ideology and was entering the service of new methods of production.

What we are now witnessing is a repetition on a new basis of the fundamental antagonism between productive forces and productive relationships which Marx with brilliant perspicacity discovered and explained.

Whilst the newly emerging proletariat elementally protested by wrecking machines and resisting inventions and science, to-day, armed with Marx', Engel's and Lenin's method of dialectical materialism, the proletariat clearly sees the path towards world freedom from exploitation of man by man.

The proletariat knows that genuine scientific knowledge of the laws of the historical process leads with irrefutable iron necessity to the conclusion that the change from one social system to another is inevitable.

The proletariat exposes all the fetishisms of class society and behind the relationships between articles sees the relationships between the human beings who create these articles.

Having learnt the real nature of the historic process the proletariat does not remain merely a spectator. It is not only the object, but the subject of the process.

The great historical significance of the method created by Marx lies in the fact that knowledge is not regarded as a passive, contemplative acceptance of reality, but as a means to effect its active reconstruction.

For the proletariat science is a means and instrument of this reconstruction. That is why we are not afraid to expose the "earthy origin" of science, and its close relations with the methods of production of material existence.

Only such a conception of science can be its real liberator from those fetters with which it is inevitably burdened in class bourgeois society.

Not only does the proletariat not fear the development of productive forces, but it alone can create all the conditions for their unprecedented development, and also for the development of science.

The teachings of Marx and Lenin have been incarnated in life. The socialist reconstruction of society is not a distant prospect, not an abstract theory, but a definite plan of great work being accomplished by the population of one-sixth of the world's globe.

And as in all epochs, in reconstructing social relationships we are reconstructing science.

The new method of research which in the persons of Bacon, Descartes and Newton, gained the victory over scholastics and led to the creation of a new science, was the result of the victory of the new methods of production over feudalism.

The building of socialism not only utilises all the achievements of human thought, but by setting science new and hitherto unknown tasks indicates new paths for its development and enriches the treasures of human knowledge by adding new treasures.

Only in socialist society will science become the genuine possession of all mankind. New paths of its development are opening before it, and there is no limit to its victorious advance, either in infinite space or in everlasting time.

THE PRESENT CRISIS IN THE MATHEMATICAL SCIENCES AND GENERAL OUTLINE FOR THEIR RECONSTRUCTION.

By E. COLMAN.

THE PRESENT CRISIS IN THE MATHEMATICAL SCIENCES AND GENERAL OUTLINES FOR THEIR RECONSTRUCTION.

By Prof. E. COLMAN.

The position of mathematics, as that of any science, is at bottom determined by the development and the position of the forces of production, of technology and economy. The latter affect mathematics both directly, by presenting it with new problems, creating its material basis and supplying its man-power, and indirectly through the prevailing outlook upon the world, the philosophy of the ruling class.

Thus, if we wish to deal with the present crisis in the mathematical sciences, we must take into consideration the crisis in bourgeois science as a whole, the crisis in the bourgeois natural sciences, and particularly in physics. The present paper, however, lays no claim to illuminate fully this aspect of the problem presented, the connection between the crisis in mathematics and the general crisis in bourgeois science, as well as its connection with the entire crisis within capitalism as a whole. It suffices to indicate that the crisis which is at the present time disturbing the whole of bourgeois science represents a new and higher phase of that crisis, experienced by physics and chemistry since the opening of the twentieth century, which Lenin analysed with the mind of a genius in his work, published in 1908—*Materialism and Empirio-Criticism.*

Of the relevant entire complex of problems, one is of special interest to us to-day: the crisis in mathematics taken in itself. Its significance is extremely great; for there is no doubt that, as the crisis in the natural sciences affects mathematics, so the crisis in mathematics, in its turn, exercises influence upon physics, chemistry and technology. On the position of mathematics, on its capacity to solve this or that problem, depends to a large extent the possibility of further development of the natural sciences and technology.

This applies with particular force to present-day physics, with its remarkably abundant mathematical apparatus, with its efforts to formalise physics, to geometricise it, its aim of allowing matter to disappear and of retaining equations only—the tendencies so strikingly revealed by Lenin. The mathematicisation of physics frightens the physicist himself; thus, for example, speaking of mathematics in physics, Khvolson says: "The most terrible thing is, that this mathematics is not the mathematics taught at the university. Not a single physicist has ever heard anything of the latter. The most terrible thing is, that its importance is continually growing." And although we oppose such obscurantism, we must admit that physics is becoming more and more dependent upon the fate of mathematics.

It is extremely characteristic that this special mathematics—the tensor analysis, the matrix calculus, the theory of characteristic numbers, has for the greater part been created by the physicists themselves, for ordinary mathematics is unable to satisfy the requirements of present-day physics.

Analysis, the most powerful instrument of mathematics, developed on the soil provided by the natural sciences, the physics, and mechanics of the 17th, 18th and 19th centuries; that is to say, on the continuity principle "*natura non facit saltus*." Mechanics and phenomenological thermodynamics regarded all processes as continuous (at least at a first approximation), and expressed them analytically. The same was true of electro-magnetism and even of chemistry.

Nevertheless, a number of discoveries were forced upon this "harmony of continuity," which, beginning with Planck's energy-quantum that was necessary to explain the heat radiation of a black body, develops through light-quanta and photons to the wave mechanics of de Broglie, Heisenberg and Schrödinger. Physics could provide no consistent analytic explanation of these phenomena and all attempts to establish a synthesis of waves and atoms, of electro-magnetism and gravitation, have so far failed, or at best, have resulted merely in *ad hoc* hypotheses. Thus, for example, Dirac's unitary theory of electricity, which is compelled to introduce the conception of negative energy, only confirms anew the contention that dialectics, which is forcing its way into science with elemental power, becomes a caricature of itself in the hands of idealist physicists. Still less working capacity is displayed by mathematics, which is scarcely able to solve the equations of wave mechanics in the most simple and trivial instances.

This inability to present an adequate picture of material reality—i.e., of the world existing independently of our consciousness—is exhibited on analysis even in comparatively elementary instances.

Let us take, for example, such a problem as that of cooling a rod, one that will appear quite simple to anybody who has not yet been corrupted by school physics and mathematics. It is well known, however, that the solution of the relevant partial differential equation with its relevant boundary conditions is of import only if we have found a series of discrete values of a parameter. Thus in itself the continuous solution of the partial differential equation does not help to elucidate the physical nature of the phenomenon. Assistance is offered by a discontinuous series of numbers and the solution breaks through the bounds of continuity.

We find similar circumstances in problems which are connected with the equilibrium of films (Plateau) and with the entire group of problems in which the given set of circumstances is dependent upon all preceding circumstances, that is, whenever we encounter integral equations, in numerous problems of variation, etc. It is obvious that this path, consisting of two wholly separate branches, along which analysis is compelled to seek a way out—first by continuous solutions, and secondly, by discrete values—can be regarded only as provisional and palliative. It is obvious that new paths must be sought.

And indeed mathematicians are trying to find a new road, but not in the right direction. As opposed to analysis, which is based upon the idea of continuity, the first quarter of our century has witnessed the vigorous development of that branch of mathematics which assigns first place to the idea of discontinuity as being self-sufficient—the theory of manifolds, the theory of real functions and, in most complete form, the theory of analytic manifolds.

Not to go too far afield, we would refer to the book by the academician Lusin (Moscow) published in Paris in 1930: *Leçons sur les ensembles analytiques et leurs applications.* The main principle depends on the formation of what is called Bair's fundamental domain. All points with rational co-ordinates are excluded from the continuum, only those points remaining which possess irrational co-ordinates; these form the fundamental domain and various operations are undertaken with them. This gives us an absolute discontinuity, for in however small a domain of any point of discontinuity you may choose, there are as many points of discontinuity of the original continuum. The question arises: where is there in material nature anything for which such an absolute discontinuity would be adequate? We find an unexpected answer to this difficult question in Lebesque's introduction to the book mentioned above: in his opinion the philosophical value of this work is even greater than its mathematical value, for its standpoint is the standpoint of solipsism. It is not difficult to understand that analytic manifolds which are built up on the concept of absolute discontinuity reflect with remarkable fidelity the extreme individualism of idealist philosophy.

We are thus presented with two extremes: continuity transformed into the absolute and the discrete made absolute, both laying claim to independent and sole significance, neither seeking reciprocal unity, each giving a one-sided picture of material reality. Consequently, the problem is not solved by rejecting both analysis and discrete mathematics as irregular and contrary to law, by dismissing both without further ado in order to start constructing afresh on clear ground; the solution lies in creating a new manner of calculation which, proceeding from a uniform principle, unites in itself the positive sides both of analysis and discrete mathematics.

The method of creating an external supplement—of discrete mathematics—to analysis, is therefore incapable of giving the synthesis of continuity and discontinuity; it is unable to lead mathematics out of its present position, in which functions containing discontinuities are treated as "unreasonable," in which gaps are considered as regrettable and undesirable revolutionary exceptions; or, on the other hand, in which the discrete is made absolute and regarded as the set of isolated individuums. We need not two analyses, but one, which in its operations would reflect the unity of continuous and discontinuous. But for such unity to be established mathematics must become dialectical and mathematicians must not fight shy of dialectical materialism.

The attempts to overcome the crisis in the present position of mathematics, which is unable to offer a synthesis of continuum and discontinuum because it is in the hands of metaphysical mathematicians, may also be described as attempts to arithmeticise analysis. The attempts are made with unsuitable means. Kronecker's old call: "Back to arithmetic!" is just as reactionary as the modern cry to let physics turn back to magic, chemistry to alchemy, and medicine to Hippocrates. We have not to arithmeticise analysis, but to introduce a new quality and bridge the gulf between continuity and discontinuity.

Apart from this gulf, however, we cannot but call attention to a second, and that is the breach between mathematics as a whole and the calculation of probability, one that is of particular importance in relation to the role of statistics in present-day physics, biology, etc.

The calculation of probability, it is well known, makes use of all branches of mathematics: arithmetic, algebra and analysis, but it is not in this that its profound connection with mathematics as a whole lies.

By the theory of geometrical probability it is known that it is possible to ascertain experimentally the value of the number π by throwing a needle upon a quadratically ruled surface and counting the number of instances in which the lines are cut by the needle, as Buffon shewed in 1777 in his *Essai d'arithmétique morale*, and as has been practically demonstrated by Wolff in Zürich.

Of the same character as this problem, in which it is possible, by a large number of experiments, to determine with a small margin of error exact mathematical magnitudes, are the well-known urn problems, such as determining the probability that, in drawing an even number of balls out of an urn containing an equal number of black and white balls, the black and white balls will appear in equal numbers, in which the expression of probability contains the square root of π Or,. as another example, the determination of the probability that, in drawing numbered balls singly out of an urn, the number of the ball drawn will not once coincide with the serial number, which gives a probability of $\frac{1}{e}$ in the case where the number of balls exceeds all limits. It may appear even more surprising that the calculation of probability enables us to find algebraic and arithmetic relations, often extremely complex. Thus, for example, Moivre's well-known problem, given in his work "De mensura sortis seu de probabilitate eventuum in ludis a casu fortuito pendetibus," published in 1711, which consists in finding the probability that, in drawing numbered balls, the total of the numbers will equal a number previously given, leads to the relation:

$$n! = \sum_{k=o}^{n} (-1)^{k} \binom{n}{k} (n-k)^{n}$$

Or, let us taken another example, quite elementary but extremely important methodologically: the summing up of a geometrical series by means of a probability procedure; this example at the same time propounds the problem of finding stochastic schemes that will enable even more complex convergent series to be added up. Finally we would also point out that the calculation of probability enables us to ascertain with sufficient accuracy such irrationals as $\sqrt{2}$ as indicated in the same connection, by Professor Khotimsky (Moscow). Actually, if we grant a margin of error of 0.5 we arrive at $2\sqrt{2} = \sqrt{8} \doteq 3$, $5\sqrt{2} = \sqrt{50} \doteq 7$, $7\sqrt{2} = \sqrt{98} \doteq 10$ from which it follows, on the theory of errors, that

$$\sqrt{2} \doteq \frac{2.3 + 5.7 + 7.10}{2^2 + 5^2 + 7^2} = 1.417$$

The general problem of determining with the desired accuracy the roots of an equation by stochastic methods can be put in analogous fashion.

The question arises how such results are possible. How is it possible that a calculation based upon chance can give exact laws, can lead to the most exact results? This question is not answered by present-day mathematics: it does not study these relations but regards them rather as a sort of curiosity, or mystery, and it is quite incapable of explaining them, because its metaphysical mode of thought is unable to grasp the unity of statistical and dynamic regularity, considering them as existing separately and antagonistically. This circumstance is, however, not only of philosophical significance; it means that present-day statistical theory is at a standstill, that the law of large numbers is not developed in the direction of varying processes. All this, in its turn, is connected with the fact that mathematics developed principally upon the basis of the natural sciences and is entirely unsuited to serve the needs of scientific political economy. It is not difficult to understand why all attempts to establish, along purely formal lines, a correspondence between the elements of the calculation of probability and of analysis have been barren of result.

The third, still broader and more profound gulf in present-day mathematics lies in the extreme lack of quality in most of its methods. By lack of quality in mathematics we do not mean, of course, that mathematical concepts have only to represent repetitions of a few main concepts, their quantitive complications. In remaining a science of magnitude, mathematics creates in magnitude itself its own specific mathematical qualities. Indeed, it develops as a whole from quantity to mass, from the abstract to the concrete, and receives new content from the influence of physics and the natural sciences generally. Nevertheless, this development is extremely contradictory: in the separate branches of mathematics it often proceeds from the concrete to greater and greater abstraction, more and more eliminating quantitative factors. An excellent example is offered by the non-Euclidean geometries. While, actually, their origin reflects a more profound knowledge of reality, in bourgeois mathematics the fact of their existence is used as conclusive evidence of the correctness of idealism and even their mathematical treatment suffers idealist distortion. In this way, that which actually forms the cognitive power of mathematics, its colossal power of abstraction at certain stages of its development, is transformed into its own contradiction and makes mathematics of a given social formation incapable of dealing with the more complicated problems that arise as our physical knowledge penetrates deeper into the profundity of matter. And even if we are not able radically to change the nature of the object of mathematics, since, on the contrary, this object determines the content of mathematics, still we must develop mathematics further in the direction of rendering more precise the qualitative variations in the sphere of quantity itself.

Mathematics reflects real dependence as functional dependence; that is to say that even the most simple dependence, the causal, is here portrayed in the most distorted, irrelevant manner; still less can mathematics portray such dependence as interaction and the unity of mutually interpenetrating contradictions. Mathematical equations have no direction, while equations which are not divested of the qualitative properties of reality—whether they are equations of the value-form in political economy, or chemical equations —possess direction by their very nature. In "Capital" Marx refers to the polarity of the simple value-form, in which the expression

20 ells of linen equals 1 coat

cannot be simply transposed into

1 coat equals 20 ells of linen.

The irreversability of chemical equations, which are exothermic in one direction and endothermic in the other, is generally known. The question then arises, why mathematical symbols, and operations with them, could not approach more closely to many of those qualitative peculiarities of reality of which mathematics would otherwise make abstractions. This way of putting the question, however, has nothing in common with the attempts to mathematicise all and sundry laws, nothing in common with the attempts to give a pan-mathematical picture of the world, attempts that rely upon Pythagoreanism and cabbalistics. The object of mathematics has limits, just like the object of physics, chemistry, biology or political economy. Just as the laws of political economy cannot be investigated by biology or physics, just as physics and chemistry are unable to elucidate biological laws, so mathematics will never be in a position to replace physics, chemistry, etc. Even if mathematics, geometry, mechanics, physics, chemistry, biology, etc., have this much in common, that they all treat of the laws of the different forms of movement of matter, and while, for example, physics and chemistry on the one hand and biology on the other cannot be regarded as so sharply separated from each other as the vitalists would have us believe, dialectical materialism does not adopt the standpoint of naive mechanistic materialism, which would reduce biology to chemistry and physics, these to mechanics and the latter to geometry and mathematics. Unity implies identity just as little as difference implies separation. Retaining its *specificum* and without laying claim to being a panacea, mathematics can and must, at a certain stage of its development on to a higher basis, turn back to its starting point, to greater concreteness, to a greater concern with quality.

This sort of attempt has already been made with success. Thus, for example, both magnitude and direction figure in the vector analysis, whose origin and development is closely bound up with the history of mechanics and electro-mechanics. Why should we not proceed further along this road and attempt to create a new qualitative calculus, for example, for the purposes of any given branch of physics? Why should we not in this way try to put into operation the grain of truth that is contained in the idealist confusion of Leibnitz, that is, to create a calculus which would be able "to represent figures and even machines and movements by letters, just as algebra represents numbers and magnitudes." (Letter from Leibnitz to Huyghens, 1679.) But to accomplish such a change in mathematics it is necessary for mathematicians to realise the true connection between concrete and abstract, to dissociate themselves from the attitude which makes a fetish of mathematics, from the idealist and non-dialectic conception of the relation between mathematics and reality.

The fourth main gulf in present-day mathematics is that between the historical and the logical. Mathematics raises problems, operates with concepts and methods that are, obviously, historically necessary, that are conditioned by the whole development of technology, the natural sciences, philosophy, and mathematics itself. But in mathematics itself, as a system of science, this historical necessity is not directly reflected. Let us take as an example the development of such a fundamental concept as that of number.

At present mathematics develops this concept on the basis, either of the so-called Hankelian permanency principle, or of Hilbert's axiomatic principle. But the permanency principle is never realised—the development of number breaks through it. As for Hilbert's axiomatics, it is true that it is of use in explaining the logical connections between individual mathematical concepts, but, since it represents a construction *post factum* it, too, is unable to give a correct picture of development. Thus, in fact, both principles only obscure the historical development that has really taken place.

We would give as a further example such an elementary fundamental expression as

$$\lim_{n \to \infty} \left(1+\frac{1}{n}\right)^n = 2.71828\ldots$$

It is well known that this expression was not artificially invented; it arose from the historical requirements of overseas trade, from the need to create, for practical calculations, the most suitable logarithmic tables. Therefore the answer to the question why this limit *e* and no other is dealt with, is given not by mathematics, but by history. Nevertheless there is some point in seeking the logical reasons for which we study precisely this limiting expression. We can also arrive at the same question by another road: we require analogous limiting expressions to be constructed, so that the number *e* is included as a link in such a series. Of course, the problem is not unambiguous, but if there were such a series it would offer us the possibility of obtaining a fuller understanding of the number *e*. Hilbert actually approaches this problem from another aspect: in the 12th point of his Paris Programme he propounds the problem of finding functions that would play the same part in any algebraic body as the exponential function plays for the body of rational numbers.

Since, in the present-day system of mathematics, there are no logical reasons why any given subject is dealt with, it seems to anybody who is not familiar with the history of mathematics that all new ideas and all new problems appear in mathematics sporadically, and to put forward problems such as those mentioned above causes confusion among mathematicians, for mathematics in its present position offers no clues to their solution. This circumstance is favourable to idealism, to *fideism*, by making it appear that it is a question of the free creation of the immanent understanding, of inspiration, and at the same time it demonstrates the incapacity of mathematics to-day to provide far-reaching generalisations that would give it a powerful impulse forward, embrace all its branches, eliminate the existing parallelism and remove from it superfluous geological stratifications. If this is to be done, the gulf between the historical and the logical in mathematics structure, methods and problematics must be revised in the light of the unity of the historical and the logical.

The fifth gulf in mathematics exists between theory and practice in mathematics itself. While mathematical theory is extremely ramified and develops very rapidly and luxuriantly, mathematical practice, that is, the methods of calculation, scarcely moves forward at all; indeed, it has practically remained stationary at the position it occupied in the 17th and 18th centuries. In actual fact, every mathematical problem, if it is to have any significance at all in practice, must eventually lead to some process of calculation, the solution which mathematical theory offers us must be effective. As it is, however, the greater part of mathematics, and particularly the more recent, gives us no effective results and scarcely troubles to do so, concerning itself rather with logical "proofs of existence."

The methods of calculation we use can in essence be traced back to the logarithms which Newton used. A large number of mathematical methods, powerful in theory, are proved in practice to be useless because the forms of their application have not been worked out. These include all self-developing processes as, for example, convergent series, products, continued fractions, iterations, recurrent processes, whose slow convergence makes them for the most part useless. Since, in present-day mathematics as in bourgeois science as a whole, "pure" mathematics is kept wholly apart from applied mathematics, the question of methods of calculation, instruments and tables that would enable whole categories of convergent series of a given type to be summed up with the required accuracy—to take one example—is not even raised, and in any case "pure" mathematics consider that it has no bearing upon their divine science. Nevertheless, the creation of such instruments would give us new logarithms which would enable us to tackle much more complicated problems than we can at present, when we are compelled to consider a problem as solved only when it leads to the main arithmetical operations. In this way the problem of reduction would be placed in quite a new light. But in order to bridge the gulf between the colossal theory and the comparatively small effectiveness of mathematics, it is essential to have a really scientific understanding of the unity of theory and practice, an understanding of the primacy of practice over theory which bourgeois mathematics to-day do not, and cannot possess.

The sixth gulf is that which exists in the so-called foundation of mathematics or, more exactly, in that philosophical superstructure whose purpose it is to justify the whole edifice, to incorporate it in the world system as a whole. The two rival philosophies for present-day bourgeois mathematics, the logistic and the intuitionist, are both idealist. It is a matter of indifference whether the world of mathematical concepts is regarded as a world of rigid immovable universals, as it is by the logists, or whether it is looked upon as the sphere of action of the free becoming as it is by the intuitionists. It is just the same if one agrees with Russell that mathematical concepts are *a priori* reducible to the natural series which is given complete in its infinite totality, or if one assumes with Weyl that mathematical concepts are born intuitively, not from sense-perception and comprehension, but in some mystic fashion apart from being. The one philosophy, just as little as the other, is able to resolve the paradoxes of the manifold theory, the law of the excluded middle, the contradiction between the singular and the many, and the finite and the infinite. The most refined finesses of finitism, of metalogic, of mathematical atomistics, merely express the anxiety of bourgeois mathematicians to separate themselves from matter and dialectics by the veil of formal logic, guiding them directly into the desert of scholasticism.

All the profound contradictions of mathematics—the contradiction between the singular and the manifold, between the finite and the infinite, the discrete and the continuous, the accidental and the necessary, the abstract and the concrete, the historical and the logical, the contradiction between theory and practice, between mathematics itself and its logical foundation—all are in reality dialectical contradictions. Their mutual inter-penetration and struggle really embody the development of mathematics. But because of the absence of plan in tackling these problems, because anarchy reigns in bourgeois mathematics as it does throughout bourgeois science as a whole and the entire capitalist system, because there is not and cannot be any general planning, the process of mathematical development encounters severe crises, experiences periods of stagnation and decline. This is reinforced by the influence of the idealist, metaphysical class philosophy and by the extreme division of labour characteristic of the whole of capitalist society and which is particularly evident in mathematics, where it is difficult to find a scientist able to understand scientific works that do not come within his own narrow province.

For mathematics there is only one way out: conscious, planned reconstruction on the basis of materialist dialectics. Is it, however, at all possible to speak of the deliberate reconstruction of mathematics and science as a whole, of the introduction of planning into science, into mathematics? The old prejudice about what is called the creative scientific work of the individual, arising from inspiration, has been destroyed by the development of science itself. There have always been scientific schools, and the creative work of one individual is rooted firmly in the collective work of the whole. And as far as planning is concerned, under capitalism the great trusts establish magnificent research institutes organised for collective work. In any case, the history of mathematics itself answers this question in the affirmative. Klein's Erlanger Programme will serve as an example; this programme, put forward in 1872, when German capitalism was growing and flourishing, accomplished a real revolution in geometry. It has had extremely fruitful results and is remarkable in that it is based upon a uniform idea, the group idea, the idea of a group of spatial transformations and of discovering the invariables which characterise it. The programme put forward by Hilbert in 1900 has been much less successful than the Erlanger Programme; it is not founded upon a uniform idea and is rather an index of separate problems—happily grasped, it is true—from all branches of mathematics. Consequently it did not accomplish in mathematics the revolution that Klein's programme did in geometry, although, despite its mosaic character, it played a positive part in directing the attention of the younger mathematicians to apparently insoluble problems—among others, Gelfond (Moscow) and Shnirelman (Moscow).

The Erlanger programme was so successful because it is not a mathematical, but a geometrical programme. Geometry, as a science that is more material than mathematics, has not only detached itself less from reality than the latter, but has frequently been of assistance to it. Geometrical methods and problems have had a wholesome effect upon mathematics by drawing it back to "sinful mother earth."

This does not imply, however, that geometry ever had more than an auxiliary part to play, that it was ever in the position to assume the leading part of the general abstract form that algebra plays in mathematics, that it could ever give more than the intuitive perception which it does give. This, too, is the standpoint from which we have to consider the most recent branch of geometry, topology, which studies the most general qualities of spatial structures that remain unchanged by any continuous uni-univalent transformations. From topology, in which a certain synthesis of the constant and the variable (group theory topology and manifold theory topology) has already been achieved and in which the school of Alexandrov, Pontriagin (Moscow) and others is making conscious efforts to give logical foundation to the fact that topological problems have been raised, a number of methods were worked out providing a solution for mathematical problems that have to be comprehended synthetically: these are the topological methods of Lusternik and Shnirelman as applied to the calculation of variations. It is to be hoped that the study of topological methods will indirectly help mathematics to progress further along this road.

If the problems enumerated above are to be solved, how is one to begin with the plan of reconstructing mathematics? The most simple thing, of course, would be if somebody should come forward with a ready-made plan, but that would be a sentence of death on the plan itself, for to-day, since mathematics has so many branches, the plan that we want can only be drawn up by collective work. Such a plan, which can be made only in a country where the national economy and science are planned, must be drawn up on the basis of all the experience gained in scientific research institutes and works' laboratories, on the basis of the requirements of the industry, agriculture and transport of the whole country. It should, however, be more than a mere register of themes suggested at conferences by scientists and engineers; it must arise as the result of intensive work on the material thus collected and its foundation must be supplied by materialist dialectics.

Hence it follows that we cannot overlook whatever has a bearing upon mathematics and its history in the writings of Marx, Engels, and Lenin. We should carefully study everything on this subject in Engels' "Anti-Dühring" and "Naturdialektik," in Lenin's "Materialism and Empirio-Criticism," in his philosophical writings and in a number of other works, particularly the economic. The hitherto unpublished writings of Marx dealing with mathematics and its history, of which there are more than fifty and which will shortly be published by the Marx-Engels Institute (Moscow), are of tremendous methodological importance.

The history of the mathematical sciences must be studied in the light of Marxist-Leninist theory, of the philosophy of dialectical materialism and the question of the tendencies in their development worked out in full detail. Above all we have to study the development of mathematics in the epoch of imperialism and the proletarian revolution, to continue upon a dialectical-materialist basis the work of Klein, the most advanced of bourgeois mathematicians, in whose works we discover, besides Machism, essential constituents of spontaneous materialism. The work of bringing out such a volume is now in progress. We have to concentrate our attention upon the development of the main mathematical concepts, particularly upon such important concepts as the differential quotient, the differential, the limiting concept. We must make a really scientific classification of the mathematical sciences; we have to investigate the importance of algebra as a connecting link between arithmetic and analysis, to study such "neglected" branches of mathematics as, for example, equations of finite differences and their connection with analysis, or divergent series; we have to find out why for the time being mathematics has turned away from them, we have to determine the position of the theory of numbers in mathematics, to study its development and the development of the entire structure of the mathematical sciences.

In addition to all these general theoretical problems there are a number of others, directly connected with socialist construction in the Soviet Union, which can be embodied into a programme of practical work for Soviet mathematicians.

We are now considering the organisation of a calculation institute for the entire Soviet Union, whose task it will be to work out new methods of calculation, to solve the mathematical problems that arise in our socialist industry, in the transport system, in collectivised and industrialised agriculture; this institute will become the operative staff for all mathematical work being conducted by the different scientific research institutions throughout the country.

One group of problems which our mathematicians have to solve is presented by the statistical problems that are so important in planned socialist economy, questions of designing and distributing economic centres (whose great importance was emphasised by Lenin), of rationalising the transport of goods, etc.

A second group is concerned with the standardisation of production, with working out optimal forms and series of products, with the problems of transmission belts and dies that arise in the rationalisation of undertakings.

The third group consists of problems that have arisen in working out the best topographical and cartographical methods corresponding to the rate of socialist construction; this is of colossal importance in a country such as the Soviet Union, with its vast natural wealth that remained uninvestigated under the predatory capitalist economy of tsarism.

There is also a fourth group of less comprehensive problems connected with the statistics of the earth's interior, with the rapidly developing building industry, the construction of airships, etc.

It is only planned socialist economy that can present mathematics with such problems, demanding the complex participation of various branches of mathematics (e.g., calculation of probability combined with differential geometry, calculation of probability and calculation of variation, or calculation of variation and the theory of numbers, etc., complex mathematical problems adjusted to mathematical mass production and maximum effectiveness, together with extensive scientific work of theoretical investigation. Thus, to plan mathematics does not mean to stifle creative work—it means that mathematics will flourish as it can do only with collective labour, the more so as the ranks of the scientists are swelled by new men from the working class, to whom the gates of the universities and research institutes have been opened wide.

The important question of reorganising the teaching of mathematics at all stages of education, from the polytechnic to the university, cannot be dealt with here.

To solve all these theoretical and practical problems, to overcome the crisis in present-day mathematics, to reconstruct it along socialist lines, patient and persistent work is necessary, the collective work of all Soviet mathematicians and of those scientists in capitalist countries who wish to go with us. Proceeding from the Leninist theory of the unity of theory and practice, we in the Soviet Union shall reconstruct the mathematical sciences. Acknowledging the Leninist principle that the sciences are not impartial, we shall place mathematics at the service of socialist construction and in this way save it from the decay that is inescapable under capitalism.

SHORT COMMUNICATION ON THE UNPUBLISHED WRITINGS OF KARL MARX DEALING WITH MATHEMATICS, THE NATURAL SCIENCES, TECHNOLOGY, AND THE HISTORY OF THESE SUBJECTS.

By E. COLMAN.

Short Communication on the Unpublished Writings of Karl Marx Dealing With Mathematics, The Natural Sciences and Technology and the History of these Subjects.

By Prof. E. COLMAN.

The following unpublished works of Karl Marx are to be found at the Marx-Engels Institute, Moscow:—

A. Extracts from works on natural science, with notes by Marx, including the following:—

1. Caspari, Otto (1878):—Leibniz' Philosophie, beleuchtet vom Gesichtespunkt der physikalischen Grundbegriffe von Kraft und Stoff.
2. Du Bois, Raymond (1878):—Leibniz'sche Gedanken in der neueren Naturwissenschaft.
3. Fick, Adolf (1876):—Die Naturkräfte in ihrer Wechselbeziehung.
4. Giordani Bruni Nolani (1851):—De triplici minime et mensure ad trium speculativarum scientiarum et multarum activarum actium principia.
5. Descartes, Renê (1878):—Opuscula posthuma physica et mathematica.
6. Poppe, J. H. M. (1853-54, 1867-68):—Geschichte der Mathematik.
7. Poppe, J. H. M. (1853-54):—Die Mechanik des 18. Jahrhunderts und der ersten Jahre des 19. Jahrhunderts.
8. Hospitalier, E. (1880-83):—Les principales applications d'électricité.
9. Allen Grant (1880-81):—Geology and History.
10. Jukes, J. B. (1878):—The Student's Manual of Geology.
11. Lyell, Ch. (1869):—Principles of Geology.
12. Fraas, C. (1878):—Klima und Pflanzenwelt in der Zeit.

13. Schleiden, M. J. (1876):—Physiologie der Pflanzen und Tiere und Theorie der Pflanzenkultur.
14. Ranke, J. (1876):—Grundzüge der Physiologie des Menschen.
15. Liebig, J. (1867):—Die organische Chemie in ihrer Anwendung auf Agrikultur und Physiologie.
16. Johnston, J. F. W. (1851):—Lectures on Agricultural Chemistry and Geology.
17. Johnston, J. F. W. (1851):—Catechism of Agricultural Chemistry and Geology.
18. Johnston, J. F. W. (1878):—Elements of Agricultural Chemistry and Geology.
19. London, J. C. (1851):—An Encyclopædia of Agriculture.
20. Liebig, J. (?):—Herr Dr. Emil Wolff in Hoferheim und die Agrikulturchemie.
21. Kaufman (1878):—Teoria praktika bankovogo dela.
22. Beckmann, J. (1860):—Beiträge zur Geschichte der Erfindungen.
23. Poppe, J. H. M. (1853-54):—Geschichte der Technologie seit der Wiederherstellung der Wissenschaften.
24. Poppe, J. H. M. (1853-54):—Lehrbuch der allgemeinen Technologie.
25. Ure, A. (1853-54):—Technisches Wörterbuch.
26. Wagner, J. R. (1878):—Die Metalle und ihre Verarbeitung.
27. Hamm, W. (?):—Die landwirtschaftlichen Geräte und Maschinen Englands.

B. Mathematical manuscripts consisting of 31 different calculations and excerpts from arithmetic, algebra, analysis and geometry, and of 19 drafts and studies for independent mathematical works. In addition, there are applications of mathematics to the problems of political economy: differential rent, the process of circulation, the rate of surplus value and the rate of profit, the problem of crises.

C. Writings dealing with technology, mostly dating from the year 1863, and concerned with the following problems:—
   1. History of mills from ancient times to the steam mill.

2. History of the weaving loom as the second main type of machine.
3. The problem of automatic production of the machine system (with the production of paper and machine construction as example).
4. The development of the tool to the machine and the machine to the system of machines.
5. The influence of mechanisation and rationalisation of production on the development of the English textile industry and on the position of the proletariat in the period 1815-1863.
6. The change in the social system of production at different stage of technical development, the relation between labour and science, between town and country, etc.